AF815412

ASTM INTERNATIONAL

SELECTED TECHNICAL PAPERS
STP1612

Editors: Norbert V. Krogstad and W. Mark McGinley

Masonry 2018

ASTM STOCK #STP1612
DOI: 10.1520/STP1612-EB

ASTM International, 100 Barr Harbor Drive, PO Box C700, West Conshohocken, PA 19428-2959
Printed in the U.S.A.

Library of Congress Cataloging-in-Publication Data

Names: Krogstad, Norbert V., editor. | McGinley, William Mark, 1958- editor. | ASTM International. Committee C07 on Lime and Limestone, sponsoring body.

Title: Masonry 2018 / Norbert V. Krogstad, W. Mark McGinley, editors.

Description: West Conshohocken, PA : ASTM International, [2018] | Series: Selected technical papers ; STP1612 | Contains peer-reviewed papers that were presented at a symposium held June 26, 2018, in San Diego, California, USA. The symposium was sponsored by ASTM International Committee C07 on Lime and Limestone. | "ASTM Stock #STP1612." | Includes bibliographical references.

Identifiers: LCCN 2018024300 (print) | LCCN 2018026484 (ebook) | ISBN 9780803176706 (ebook) | ISBN 9780803176690 (pbk.)

Subjects: LCSH: Masonry--Congresses.

Classification: LCC TH5317 (ebook) | LCC TH5317 .M36957 2018 (print) | DDC 620.1/3--dc23

LC record available at https://lccn.loc.gov/2018024300

ISBN: 978-0-8031-7669-0

Peer Review Policy

Each paper published in this volume was evaluated by two peer reviewers and at least one editor. The authors addressed all of the reviewers' comments to the satisfaction of both the technical editor(s) and the ASTM International Committee on Publications.

The quality of the papers in this publication reflects not only the obvious efforts of the authors and the technical editor(s), but also the work of the peer reviewers. In keeping with long-standing publication practices, ASTM International maintains the anonymity of the peer reviewers. The ASTM International Committee on Publications acknowledges with appreciation their dedication and contribution of time and effort on behalf of ASTM International.

Citation of Papers

When citing papers from this publication, the appropriate citation includes the paper authors, "paper title," STP title, STP number, book editor(s), ASTM International, West Conshohocken, PA, year, page range, paper doi, listed in the footnote of the paper. A citation is provided on page one of each paper.

Printed in Mayfield, PA
September, 2018

Foreword

THIS COMPILATION OF Selected Technical Papers, STP1612, *Masonry 2018*, contains peer-reviewed papers that were presented at a symposium held June 26, 2018, in San Diego, California, USA. The symposium was sponsored by ASTM International Committee C07 on Lime and Limestone.

Symposium Chairs and STP Editors:

Norbert V. Krogstad
Wiss, Janney, Elstner Associates, Inc.
Northbrook, IL, USA

W. Mark McGinley
University of Louisville
Louisville, KY, USA

Contents

Overview

These are the proceedings of the ASTM symposium entitled *Masonry 2018: Innovations in Collaborative Research, Development, and Applications* held on June 26, 2018, in San Diego, California. The purpose of this symposium was to gather and disseminate the latest information on innovations in masonry materials, design, specification, construction, maintenance, and rehabilitation. It specifically addressed the application of ASTM standards to these areas and their coordination with building codes, project specifications, and international standards.

The 11 papers presented in this STP were all peer reviewed. The symposium and resulting STP is cosponsored by Committees C01 on Cement, C07 on Lime and Limestone, C12 on Mortars and Grouts for Unit Masonry, and C15 on Manufactured Masonry Units. The authors of the papers represent a wide range of constituents, including masonry consultants, industry groups, and academics from both US and foreign universities. As an example of this range, there are authors from the Brick Industry Association, the National Brick Research Center at Clemson University, the New Mexico Institute of Mining and Technology, and the Bangladesh University of Engineering and Technology.

The symposium and STP are divided into three sessions/sections. The first session/section is entitled *Measurements and Testing*. The papers in this section include research to determine moisture content of masonry for use in analytical hygrothermal models, studies to develop an alternate approach to calculate normal stress in brick masonry for in-situ shear tests, and a study to identify dominant natural periods of variable story unreinforced masonry buildings in Dhaka.

The second session/section is entitled *Materials and Performance*. This section features papers describing the development of new masonry units made from cellulose and bentonite clay that can be modified using woodworking tools; development and calibration of a new modified hotbox device for measuring thermal performance of nonhomogeneous masonry materials; and a study that looked at 24-hour thermal loading in a modified hotbox to measure thermal mass effects of masonry systems.

The last session/section is entitled *Design and Code*. This section includes papers that discuss aesthetic issues of buildings constructed in Chicago, IL, USA from 1932 through 2017, code requirements for stone masonry veneers, and recommendations for code modifications to support use of stone veneer cavity walls; the use of glass fiber polymer flexural reinforcement to reinforce beams (with recommended code changes); and two papers that address cold formed steel stud back masonry veneer wall systems, both the history of development and recommended code changes to improve wall system performance.

The symposium and STP are a result of years of work by the authors of the papers, ASTM staff members, and volunteer reviewers. The joint symposium committee was comprised of Bill Kjorlein and Jamie Farny from ASTM C01; Mike Tate from ASTM C07; Brian Trimble and Laura Powers of ASTM C12; and was chaired by Norbert Krogstad and Dr. Mark McGinley of ASTM C15.

The papers presented in this STP further advance our knowledge of masonry and its systems with the aim of improving the use in building construction in both the US and throughout the world.

STP 1612, 2018 / available online at www.astm.org / doi: 10.1520/STP161220170167

Rex A. Cyphers,[1] Annette M. Cyphers,[2]
Jodi M. Knorowski,[1] and Ashley M. Skertic[3]

Evaluation of Existing Moisture Content in Brick for Hygrothermal Models in the Rehabilitation of Mass Masonry Walls

Citation

Cyphers, R. A., Cyphers, A. M., Knorowski, J. M., and Skertic, A. M., "Evaluation of Existing Moisture Content in Brick for Hygrothermal Models in the Rehabilitation of Mass Masonry Walls," *Masonry 2018, ASTM STP1612*, N. V. Krogstad and W. M. McGinley, Eds., ASTM International, West Conshohocken, PA, 2018, pp. 1–12, http://dx.doi.org/10.1520/STP161220170167[4]

ABSTRACT

A research study was undertaken to evaluate an in situ test method for determining the existing moisture content of brick masonry for use in hygrothermal models. Currently, ASTM E3069, *Standard Guide for Evaluation and Rehabilitation of Mass Masonry Walls for Changes to Thermal and Moisture Properties of the Wall* [1], is one of the only industry references providing guidance for evaluating the existing moisture content of masonry materials in buildings when the addition of insulation or other changes to the wall assembly are being considered. When interior insulation or a continuous air barrier is considered for addition to mass masonry walls to meet new energy standards, understanding the properties of the brick, the condition of the wall, and the existing moisture content of the brick are imperative for evaluating the hygrothermal performance of the wall assembly. The assumptions made for the initial moisture content of materials in the hygrothermal models for a mass

Manuscript received September 15, 2017; accepted for publication December 18, 2017.
[1]WDP & Associates Consulting Engineers, Inc., 335 Greenbrier Dr., Suite 205, Charlottesville, VA 22901, USA
R. A. C. http://orcid.org/0000-0003-0034-7840, J. M. K. http://orcid.org/0000-0002-9170-6839
[2]University of Virginia, 575 Alderman Rd., Charlottesville, VA 22904, USA http://orcid.org/0000-0003-2295-3307
[3]West Virginia University Statler College of Engineering, Morgantown, WV, USA http://orcid.org/0000-0001-7626-9928
[4]ASTM International Symposium on *Masonry 2018* on June 26, 2018 in San Diego, CA, USA.

masonry wall can change the design and alteration approach and, in some cases, could lead to a repair that would have detrimental long-term effects. In the laboratory, the interstitial relative humidity within the brick was measured utilizing temperature and humidity probes embedded within the brick body. The moisture content of the brick was measured at various humidity levels and then compared to various published isotherm curves. In addition to the laboratory testing, field-testing was conducted at five historic buildings with mass masonry wall assemblies to assess the moisture content of the masonry utilizing the same methods as were used for the laboratory testing. When within the hygroscopic range, these methods could be used as minimally invasive and direct techniques to measure the existing moisture content of the brick to develop more accurate and situationally specific hygrothermal models when evaluating existing mass masonry walls prior to a renovation.

Keywords

in situ relative humidity, moisture content, historic mass masonry, sorption isotherm curve

Introduction

When performing a hygrothermal analysis, the assumptions made regarding material properties and initial conditions can significantly alter the results of the analysis. To improve the accuracy of hygrothermal models, a complete understanding of the unique conditions of a given wall assembly and material components is important. Criteria for the inputs and evaluation of outputs are provided in the American Society of Heating, Refrigeration, and Air Conditioning Engineers (ASHRAE) Standard 160 [2]. Laboratory testing can be performed on material components to determine the material properties necessary to perform such an analysis. While many material properties are static, the moisture content of the materials will vary over time as surrounding conditions change. As such, methods for determining the initial moisture content should consider this dynamic effect.

Sorption isotherm curves show the relationship between relative humidity and moisture content of a material at a constant temperature. They are developed by incrementally exposing a material to increasing or decreasing levels of moisture and measuring the actual moisture content at each increment once the conditions have normalized in accordance with ASTM C1498, *Standard Test Method for Hygroscopic Sorption Isotherms of Building Materials* [3]. When exposed to increasing levels of relative humidity, the adsorption properties of the material are measured. The desorptive properties are measured when the material is exposed to decreasing levels of relative humidity, starting from a point when the material is fully saturated. Sorption isotherm curves for brick show the moisture content of the brick remains generally unchanged until exposed to relative humidity levels above 90 %. In a mass masonry wall, this behavior is depended upon to manage the expected amount of moisture.

An understanding of sorption isotherm curves is important when evaluating the hygrothermal behavior of mass masonry walls because the conditions that the materials in the wall are exposed to are constantly changing, and as such, the material is always in a state of either taking on or releasing moisture. Properties such as porosity and capillary suction factor in as to how the brick is impacted by bulk water, but the movement of water vapor through the brick must also be evaluated for long-term performance of an overall wall assembly.

The initial moisture content is the ratio of the mass of water in the material to the mass of solids in the same material prior to any exposure to other substances. From a hygrothermal analysis standpoint, the initial moisture content is the moisture content of the materials in the wall assembly at the start of a model simulation. This factor is important and can impact the placement and selection of insulation or other vapor-resistive materials designed to be part of the wall assembly in a retrofit application. If the initial moisture content is relatively high, the wall assembly must be designed to allow the moisture to dry out and dissipate without causing any long-term damage to the materials within the wall assembly that can be impacted by elevated moisture levels or cause a significant reduction in thermal resistance of the insulation. For lower initial moisture contents, there is less potential for moisture to become trapped in the wall as a result of the installation of new materials.

When evaluating an existing masonry wall assembly, determining the initial moisture content generally is not feasible without removing a sample of brick, sealing it in an air- and vapor-tight bag, and taking it to a laboratory for testing. This process can be invasive and disruptive to the existing building, especially if it has historical significance. As previously noted, the moisture content is also constantly changing based on the surrounding conditions, so this method only provides a relatively small data sample of the actual moisture content. The measurement of initial moisture content utilizing in situ data logging instrumentation provides a less invasive and more efficient way of gathering data. Furthermore, the overall movement of moisture within a wall assembly can be monitored over time and include effects of weather events or unique climatic conditions along with this more accurate representation of the moisture content.

Testing Procedures

For this study, new samples of five whole hand-molded and five whole wood-molded brick were selected for testing. These types of brick were selected because they are representative of brick typically found in historic mass masonry walls. Testing data provided by the manufacturer for each type of brick are provided in **Table 1**.

The ten whole bricks that were selected were generally square and true to size and shape without significant spalls or defects. A core was drilled at the center of each brick sample to a depth of approximately one-half the depth of the brick as

TABLE 1 Testing data provided from manufacturer.

Test	Hand-Molded Brick	Wood-Molded Brick
Absorption: 24-H Submersion in Cold Water	8.92 %	6.30 %
Absorption: 5-H Submersion in Boiling Water	14.38 %	9.84 %
Saturation Coefficient	0.61	0.64
Compressive Strength	6,116 psi	7,375 psi
Efflorescence	Not Effloresced	Not Effloresced
IRA (Oven Dried Method)	45.8 g/min/30 in^2	23.9 g/min/30 in^2
Abrasion Resistance	0.15	0.09

shown in **Fig. 1**. The hole was sized to suit the installation of an impermeable liner and the interstitial probe.

Each of the ten whole brick samples was placed into a heated oven to determine the dry weight in accordance with ASTM C67, *Standard Test Methods for Sampling and Testing Brick and Structural Clay Tile* [4]. Once the weight was obtained, an impermeable liner was then placed into the core and sealed to the surface of the brick in an air- and a vapor-tight manner. A relative humidity and temperature probe was then inserted into the liner, and the open end was sealed to prevent air infiltration. The entire assembly was then weighed to obtain the dry weight of each sample including the instrumentation. This was performed so the actual moisture content could be calculated for each relative humidity level. The assembly was then placed in an environmental chamber and exposed to preset amounts of relative humidity and a constant temperature. The temperature remained at a constant level throughout testing of variable humidity levels. The ten whole brick samples were exposed to relative humidity levels of approximately 60 % and 80 % in the environmental chamber and were then moved to a fog room with relative humidity levels consistently near 100 %. The temperature was maintained at between 70°F and 75°F during each of these tests. **Fig. 2** shows the instrumented, labeled, wood-molded brick samples inside the environmental chamber. An ambient temperature and relative humidity data logger was also placed within the chamber to verify the ambient conditions.

FIG. 1 Location of hole drilled into each brick sample.

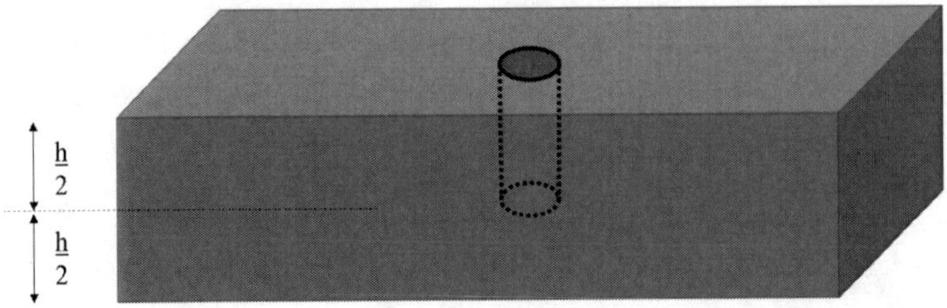

FIG. 2 Brick samples with interstitial probes installed inside environmental chamber.

Once the moisture content of the brick was determined, it was then correlated with the in situ relative humidity value measured by the probes.

Results of Testing

LABORATORY TESTING

The density of the sample bricks was used to compare the results of the samples tested to published values. Table 2 shows the density of each brick sample for both the hand-molded and wood-molded bricks that were tested as part of this study. Three routinely referenced publications contain sorption isotherm data for different types of brick: WUFI, a hygrothermal modeling software, provides approximate data for different types of brick [5]; ASTM MNL40, *Moisture Analysis and Condensation Control in Building Envelopes* [6]; and ASTM MNL18, *Moisture Control in Buildings* [7]. The density of different types of brick referenced in these publications is presented in Table 3. The published sorption isotherm curves for each of these references is shown in Fig. 3.

TABLE 2 Density of brick samples.

Sample Label Number	Hand-Molded (lb/ft^3)	Wood-Molded (lb/ft^3)
1	105.8	115.4
2	107.4	123.9
3	109.6	124.1
4	108.8	120.7
5	108.4	125.3
Average	108.0	121.9

TABLE 3 Published density values of brick.

Publication	Density (lb/ft^3)
WUFI (Hand-Molded) [5]	107.7
WUFI (Historical) [5]	112.4
WUFI (Solid) [5]	118.6
ASTM Manual 40 (Red Brick) [6]	104.3
ASTM Manual 18 (Reheated Red Brick) [7]	112.4

FIG. 3 Published sorption isotherm curve, data from Refs. [5–7].

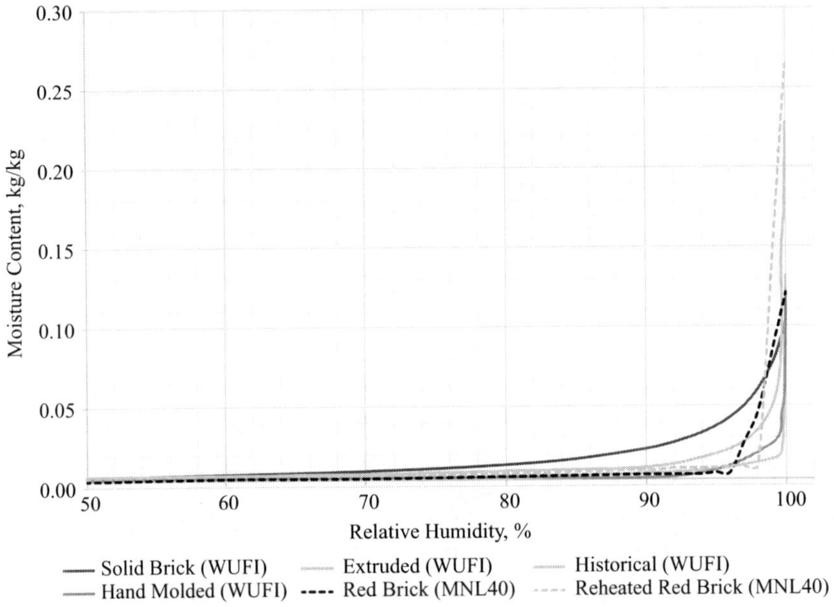

The data collected from the relative humidity probes were analyzed in several ways. First, the in situ relative humidity for each brick sample was compared to the ambient relative humidity within the environmental chamber (**Fig. 4**). Initially, the in situ relative humidity for the brick samples was slightly higher than that of the ambient relative humidity. While the ten brick samples had been oven-dried to determine their dry weight, they remained at room temperature for a number of hours while the probes were installed and sealed. Based on the initial in situ readings, it appears the brick samples absorbed moisture from the air during this time and then had to dry back out when placed in the environmental chamber at a lower relative humidity. When the relative humidity was increased to 77 %, there was a

FIG. 4 In situ relative humidity versus ambient relative humidity.

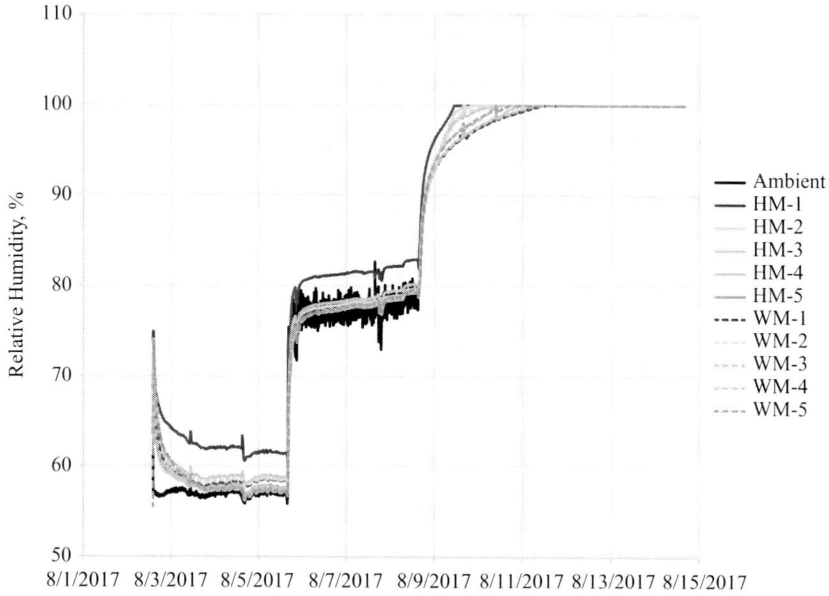

slight lag in the in situ readings before they normalized and remained relatively consistent with the ambient conditions. When the samples were moved to the fog room, the ambient logger could not monitor the conditions without being damaged by the high relative humidity levels, but the fog room is consistently held at a relative humidity near 100 %. The in situ readings took several days to normalize to the ambient conditions, but they all eventually reached the 100 % threshold.

Partial sorption isotherm curves were then developed for each of the brick samples as shown in **Fig. 5**. Generally, the hand-molded brick was found to adsorb more moisture than the wood-molded brick at the same relative humidity levels. The sorption isotherm curves for each type of brick were then compared to the published data for bricks with similar densities. These comparisons are shown in **Fig. 6** for hand-molded brick and **Fig. 7** for wood-molded brick.

FIELD-TESTING

In addition to the laboratory testing, field-testing utilizing the same in situ probes was performed at five historic buildings with mass masonry walls. For existing wall assemblies, ASHRAE Standard 160 states that the initial moisture content should be modeled at a level equivalent to the moisture content of the material at 90 % relative humidity for concrete and 80 % for all other materials [2]. This field-testing was intended to measure actual relative humidity and moisture content for a representative selection of brick in mass masonry walls for comparison to the standard guidance.

FIG. 5 Sorption isotherm curves for brick samples.

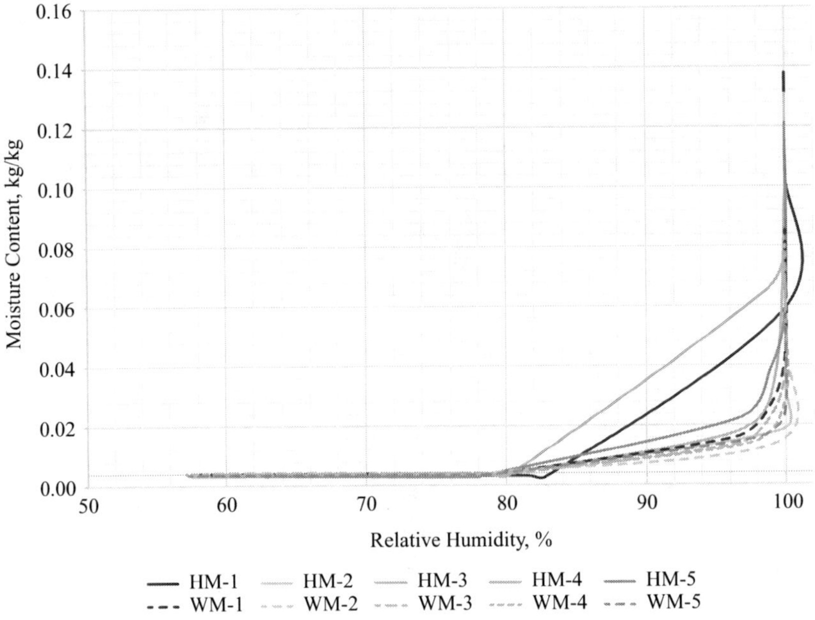

FIG. 6 Published sorption isotherm curve compared to in situ readings for hand-molded brick, published data from Refs. [5,6].

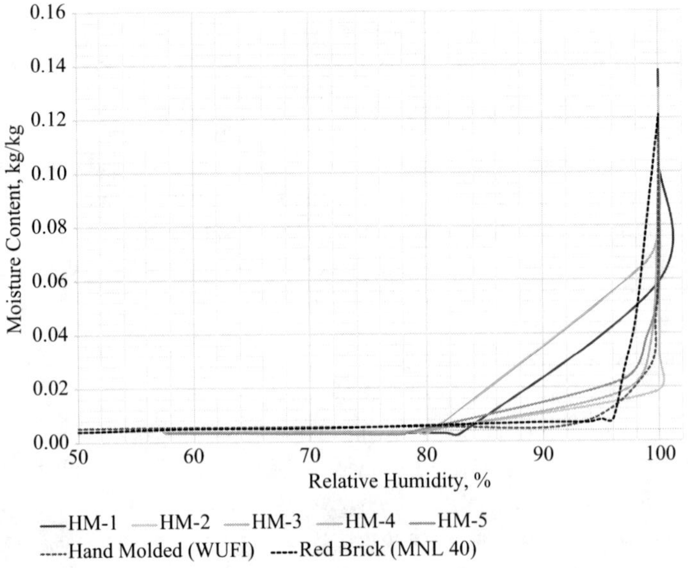

FIG. 7 Published sorption isotherm curve compared to in situ readings for wood-molded brick, published data from Ref. [5].

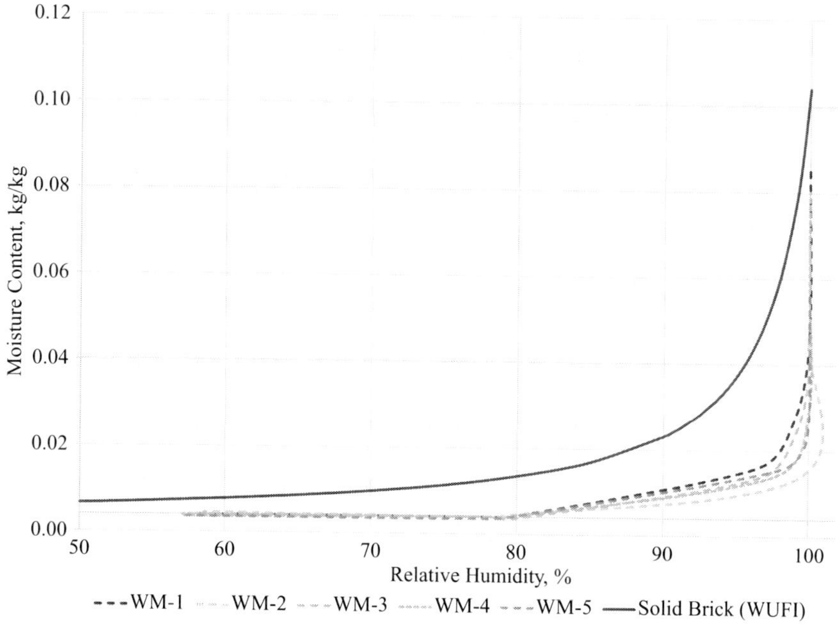

The thicknesses of the walls and materials varied at each of the five buildings and, in all cases, met the definition of a mass masonry wall [1]. These walls rely on the moisture storage capacity of the materials to prevent bulk water leakage with no flashings or other drainage provision present. The buildings were all constructed between 60 and almost 200 years ago. Based on visual observations, Buildings 1 through 4 all appeared to be in good condition with no signs of active or previous water infiltration. Building 5 appeared to be in good condition based on exterior observations, but signs of previous moisture damage were present on the interior of the building. The wall assemblies in each of the five buildings included at least one wythe of brick similar to the units selected for the laboratory testing.

The areas tested on Building 1 were typically comprised of four to five brick wythes, an air space, interior clay tile, and had exterior granite in one area. The areas tested on Building 2 were comprised of two to three brick wythes and an air space and were finished on the interior with traditional plaster. In addition, one area on Building 2 was retrofitted with rigid insulation. The areas tested on Building 3 were comprised of two brick wythes, an air space, and interior hollow clay tile. The areas tested on Building 4 were comprised of concrete masonry backup with brick exterior. The areas tested on Building 5 were typically comprised of three brick wythes and exterior Indiana limestone.

Probes to log temperature and humidity were placed into the walls from the interior of the building, taking care to not place the probe too close to the exterior surface and risk damage to the exposed face of the wall. Similar to the probe installation for the laboratory testing, an impermeable liner was installed around the probes. The probes were left in place for as long as possible based on user requirements for the buildings. At a minimum, they were left in place long enough to accumulate data during and after weather events to evaluate the impact of the moisture and the movement of the moisture through the wall assembly. Probes in Buildings 1, 2, 3, and 5 remained installed for a period of months. Probes in Building 4 were left in place for one week.

For the purpose of this discussion, the bricks tested have been identified as interior, middle, or exterior wythes based on their relative location in the wall assembly. When in situ probes were installed in well-constructed wall assemblies found in Buildings 1 through 4, the relative humidity values were often found to be significantly less than the design equivalent moisture content relative humidity (**Fig. 8**). The average relative humidity among the interior brick wythes ranged from 52.5 % to 64.9 %. The average relative humidity among the middle brick wythes ranged from 46.9 % to 69.5 %. The average relative humidity among the exterior brick wythes ranged from 64.3 % to 85.8 %.

FIG. 8 Probe data, Buildings 1 through 4.

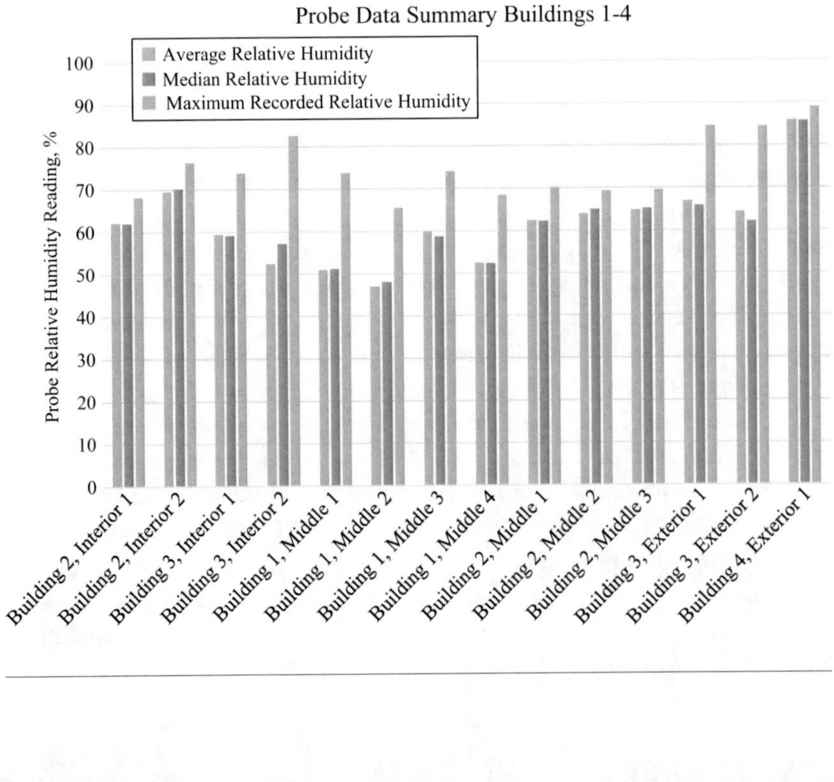

While there were previous signs of moisture damage at Building 5, there were no reports of active water infiltration. The initial interstitial relative humidity readings typically exceeded 90 %. Shortly after the probes were installed, a significant rain event occurred, and the interstitial relative humidity readings went up to 100 % and stayed at this level until they were removed. This is indicative of active water leakage that was confirmed by subsequent field-testing that is beyond the scope of this paper.

Conclusions

The results of the laboratory testing showed that interstitial relative humidity probes could reliably be used to estimate the moisture content of brick in the hygroscopic range. Published and available sorption isotherm curves can be used to provide a reasonable estimate of moisture content if it is not feasible to develop a unique sorption isotherm curve for hand- or wood-molded bricks. The study showed that the use of interstitial relative humidity probes in mass masonry walls is a viable method to develop more accurate hygrothermal models in a direct and minimally invasive manner.

Careful consideration should be given to determining the existing moisture content of the materials within mass walls when undertaking hygrothermal modeling or planning to add interior insulation. The study showed that there was variation in the moisture content of the different wythes within the same wall. However, the moisture content of interior wythes in wall assemblies in good condition remained relatively dry and was not impacted by weather events. The interstitial relative humidity in the interior wythes remained consistently below 80 %, with only one wall assembly reaching 82 % for a short period of time. Conversely, the moisture content of the wall assembly with interior leakage was high and remained high for a long period of time. The maximum interstitial relative humidity was measured to be above 80 % for the walls determined to be in good condition. Applying a moisture content equal to the equilibrium moisture content at 80 % could, in some cases, underestimate the moisture content and, in others, would overestimate the amount of moisture within the materials. The results of the field study also showed that the use of interstitial humidity probes could be used as a diagnostic tool to assist in determining the presence of moisture-related problems.

Future Studies

A similar laboratory and field study is currently planned to be conducted on mortar samples and on prisms constructed on the same mortar and the bricks used for this study. It is anticipated that laboratory testing will also be expanded to include extruded bricks. In situ probes will be installed at both brick and mortar locations within the prism, and the assembly will be subjected to the same relative humidity levels. The goal of this study is to see how the interstitial readings from the assembly correlate to the readings obtained from the individual components of the assembly. In a field application, specifically in a multi-wythe wall, it may be difficult to identify whether the probes are being installed into brick or mortar at various depths within the assembly.

References

[1] ASTM E3069-17, *Standard Guide for Evaluation and Rehabilitation of Mass Masonry Walls for Changes to Thermal and Moisture Properties of the Wall*, ASTM International, West Conshohocken, PA, 2017, www.astm.org

[2] ASHRAE 160-16, *Criteria for Moisture-Control Design Analysis in Buildings*, ASHRAE Inc., Atlanta, GA, 2016, www.ashrae.org

[3] ASTM C1498-04, *Standard Test Method for Hygroscopic Sorption Isotherms of Building Materials*, ASTM International, West Conshohocken, PA, 2004, www.astm.org

[4] ASTM C67-17, *Standard Test Methods for Sampling and Testing Brick and Structural Clay Tile*, ASTM International, West Conshohocken, PA, 2004, www.astm.org

[5] "WUFI Pro, Version 6.1," Fraunhofer Institut Für Bauphysik, Stuttgart, Germany, 2017, www.wufi.com

[6] Trechsel, H. J., Ed., *Moisture Analysis and Condensation Control in Building Envelopes, MNL40*, ASTM International, West Conshohocken PA, 2001

[7] Trechsel, H. J., Ed., *Moisture Control in Buildings: The Key Factor in Mold Prevention, 2nd ed., MNL18*, ASTM International, West Conshohocken, PA, 2009.

STP 1612, 2018 / available online at www.astm.org / doi: 10.1520/STP161220170181

Francesco Graziotti,[1] Gabriele Guerrini,[1] Andrea Rossi,[2]
Guido Andreotti,[1] and Guido Magenes[1]

Proposal for an Improved Procedure and Interpretation of ASTM C1531 for the In Situ Determination of Brick-Masonry Shear Strength

Citation

Graziotti, F., Guerrini, G., Rossi, A., Andreotti, G., and Magenes, G., "Proposal for an Improved Procedure and Interpretation of ASTM C1531 for the In Situ Determination of Brick-Masonry Shear Strength," *Masonry 2018, ASTM STP1612*, N. V. Krogstad and W. M. McGinley, Eds., ASTM International, West Conshohocken, PA, 2018, pp. 13–33, http://dx.doi.org/10.1520/STP161220170181[3]

ABSTRACT

In situ evaluation of the mechanical properties of existing masonry structures can be a challenging problem because destructive in situ tests are not always feasible. Cohesion and friction coefficients of the mortar-brick interface are fundamental parameters to characterize the shear strength of brick-masonry walls: the slightly invasive in situ test called the "shove test" can be performed to establish rational values for these parameters as described in ASTM C1531, *Standard Test Methods for In Situ Measurement of Masonry Mortar Joint Shear Strength Index*. During the shove test, a jack pushes one brick of the wall horizontally, while two flatjacks impose a known vertical stress on the masonry portion surrounding the selected unit. In order to insert the horizontal jack and allow the brick to slide, the two adjacent bricks need to be removed, introducing discontinuities within the compression stress flow. At the same time, the service load acting on the pier

Manuscript received September 29, 2017; accepted for publication December 14, 2017.
[1]Dept. of Civil Engineering and Architecture, University of Pavia - EUCENTRE, Via Ferrata 3, 27100 Pavia (PV), Italy, F. G. http://orcid.org/0000-0002-0223-0139, G. G. http://orcid.org/0000-0001-7853-3710, G. A. http://orcid.org/0000-0003-2649-9631, G. M. http://orcid.org/0000-0002-5452-1501
[2]UME Graduate School, University School for Advanced Studies IUSS, Piazza della Vittoria 15, 27100 Pavia (PV), Italy http://orcid.org/0000-0002-7119-1778
[3]ASTM International Symposium on *Masonry 2018* on June 26, 2018 in San Diego, CA, USA.

imposes an additional compressive stress on the bed-joint interface. Moreover, sliding of the test brick would require vertical expansion of the restrained bed joint, resulting in further compression due to dilatancy. Knowing the actual stress distribution after removal of the two bricks and considering the service-load and dilatancy effects are essential to determine the actual compressive stress on the test unit and to derive the correct shear strength parameters from the force applied by the horizontal jack. However, current standards do not address these effects explicitly. The work presented in this paper, validated through in situ and laboratory tests and detailed numerical analyses, consists of an improved methodology for the execution and interpretation of the shove test that estimates the actual compressive stress on the selected brick and consequently provides a more rational interpretation of the shear resistance of the bed joints.

Keywords
shove test, shear strength, unreinforced brick masonry, in situ test, dilatancy

Introduction

Evaluating the structural performance of existing brick masonry buildings requires knowledge of their actual material properties, a task that involves consideration of complex mechanical interactions among the constituents of masonry. In fact, bricks and mortar are two independent materials, characterized by inherently different properties; however, they work together in resisting the applied loads, constituting a composite masonry material. Consequently, determining only the mechanical properties of each individual constituent may not be sufficient because it is necessary to characterize the behavior of masonry as an overall composite material.

The shear strength of masonry is one of the fundamental parameters necessary for the evaluation of unreinforced masonry structures, especially when their seismic response is under examination. Among several techniques currently available to determine this parameter, the "shove test" described in ASTM C1531-16, *Standard Test Methods for In Situ Measurement of Masonry Mortar Joint Shear Strength Index* [1], appears particularly suitable for regular brick masonry in case of through-joints failure. This in situ test is mildly invasive and allows easy restoration of the tested wall once the procedure is terminated.

The shove test consists of isolating a brick unit by removing the two adjacent bricks and pushing (shoving) it with a horizontal jack. A controlled vertical stress is imposed on the surrounding masonry by two flatjacks located above and below the selected brick. The procedure can be repeated under different vertical stresses. A Coulomb-type frictional relationship finally is derived between the vertical compressive stress and the shear strength index of the bed-joint interfaces and is characterized by cohesion and friction coefficients.

Comparing the Coulomb parameters determined through laboratory shove tests with those obtained via triplet tests [2,3] may result in significant differences [4]. It has been observed that this discrepancy may be caused by an incorrect estimation of the normal compressive stress acting on the test brick or, equivalently, on the brick-mortar interface. In particular, determining the actual compressive stress induced by the flatjacks on the test unit after removal of the two adjacent units is of fundamental importance. The normal compressive stress also may include a non-negligible far-field contribution due to the overburden stress acting on the surrounding masonry. Moreover, sliding of the test brick would require vertical expansion of the restrained bed joint, resulting in further compression due to dilatancy. ASTM C1531-16 [1] addresses these issues by proposing simplified estimates of the actual normal compressive stress on the test unit that are valid only for specific geometric conditions and whose general accuracy could be questioned.

To this end, the present work proposes a method to evaluate the actual compressive stress acting on the tested brick, considering the three effects mentioned previously: (i) variation of the stress field due to removal of the adjacent bricks, (ii) far-field effect, and (iii) dilatancy. Once the normal compressive stress is determined, the results of the shove test can be interpreted more realistically, thus improving the accuracy of the frictional relationship between shear strength index and compressive stress of the masonry.

ASTM C1531-16 Testing Procedure and Interpretation

The purpose of the test is to estimate the parameters necessary to compute the shear strength of brick masonry in case of through-joints failure (i.e., strong brick and weak mortar), evaluating in situ the shear strength index of the two horizontal bed joints bounding a single brick unit. This type of failure usually is represented by Coulomb-type criteria [5]. This formulation is based on two parameters: the cohesion, τ_0 (i.e., the shear strength without normal compression), and the friction coefficient, μ. ASTM C1531-16 [1] outlines three possible test procedures. "Method A," which controls the vertical stress with flatjacks above and below the tested brick and applies the horizontal load by a small cylindrical jack (**Fig. 1**), appears to be the most complete.

FIG. 1 ASTM C1531-16 shove test setup [1].

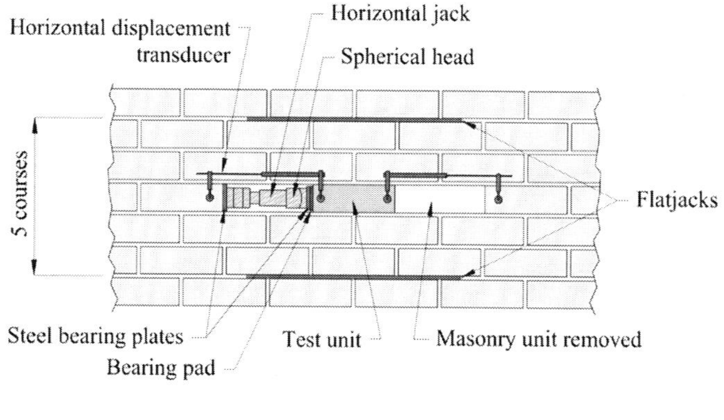

METHOD A STEP-BY-STEP TESTING PROCEDURE

Method A of ASTM C1531-16 [1], "Procedure with Flatjacks Controlling Normal Compressive Stress," can be summarized in the following step-by-step procedure:

1. Identify the brick unit to be tested, which must be in the stretcher orientation; locate the test site far enough from wall openings or ends so that the loading jack bears against enough masonry to resist forces generated during loading of the test unit.
2. Prepare two slots in the masonry bed joints, two courses above and two courses below the test unit; insert the flatjacks in the two slots, ensuring they are parallel to each other and vertically aligned with the test brick, according to ASTM C1197-14a, *Standard Test Method for In Situ Measurement of Masonry Deformability Properties Using the Flatjack Method* [6].
3. Remove the two bricks adjacent to the test unit.
4. Instrument the test unit with one or two horizontal displacement transducers; place the loading jack, spherical seat, and bearing plates with bearing pads into the space next to the test unit; shim the jack to provide a centered horizontal force on the test unit (**Fig. 1**).
5. Set the flatjack pressure equal or less than 0.07 MPa (10 psi) and hold it constant; increase pressure in the horizontal loading jack gradually, recording displacement of the unit at small increments of horizontal load; when the bond between the unit and mortar joints is broken, the unit will begin to displace continuously under constant horizontal load; this represents the maximum horizontal load for this first level of normal compressive stress.
6. Increase the pressure in the flatjacks to the next desired value and repeat the process of horizontal loading until the maximum horizontal load for this level of normal compressive stress is reached; repeat the sequence in this manner to determine the bed-joint residual shear stress at various levels of normal compressive stress.
7. To investigate the effect of shear force reversal on the in situ shear strength, if desired, the horizontal jack may be transferred to the opening on the opposite side of the test unit, the displacement instrumentation reversed, and the test sequence repeated in the opposite direction.
8. Release pressure in the horizontal jack after the final displacement measurement has been taken and remove the loading jack; release pressure in the flatjacks and remove the flatjacks and displacement measurement devices.

DATA INTERPRETATION

The output of the test consists of displacement transducer and pressure gauge recordings at each increment of vertical compression. For a specific flatjack pressure, it is possible to plot the pressure of the horizontal jack versus the horizontal displacement of the test unit, identifying the point corresponding to bond failure. This is the point associated with the peak horizontal jack pressure, immediately preceding its sudden drop when the brick begins to slide. The peak shear strength index, τ_c, corresponding to the peak horizontal force right before the onset of sliding, P_{hc}, is given by Eq 1:

$$\tau_c = \frac{P_{hc}}{A_j} \tag{1}$$

where A_j is the sum of the gross areas of the upper and lower mortar bed joints. If a double-wythe wall is tested, it is recommended to remove the vertical mortar joint behind the test unit before the test execution.

However, the peak shear strength index, τ_c, is implicitly discarded in Method A where only forces measured during brick sliding are considered, as stated in ASTM C1531-16 §9.1.2: "When the bond between the unit and mortar joints is broken, the unit will begin to displace continuously under constant horizontal load. This represents the maximum horizontal load for this level of normal compressive stress" [1]. Consequently, only the residual horizontal forces, measured during sliding of the test brick, $P_{h,i}$, are taken into account to calculate the mortar joint shear strength index at any i^{th} level of flatjack pressure (**Fig. 2**) using Eq 2:

$$\tau_i = \frac{P_{h,i}}{A_j} \tag{2}$$

At the ith level of flatjack pressure, the vertical stress applied by the flatjacks to the masonry in contact with them, $\sigma_{j,i}$ (**Fig. 2**), is recorded in accordance with ASTM C1197-14a [6]. However, the relationship between joint shear strength index and compressive stress in the masonry in contact with the flatjacks, σ_j, does not give direct information on the actual frictional shear strength of the masonry. In fact, the most useful information is the relationship between the joint shear strength index and the effective normal compression in the test unit brick, σ_b (**Fig. 2**). Appendix X1 of ASTM C1531-16 [1] suggests that the average normal stress on the test unit at the ith level of flatjack pressure, $\sigma_{b,i(ASTM)}$, can be approximated as (Eq 3):

$$\sigma_{b,i(ASTM)} = j \cdot \sigma_{j,i} \tag{3}$$

where j is a modification factor accounting for normal stress concentration on the test unit due to removal of the adjacent units; a value of 1.7 is proposed for the test configuration presented as an example in the standard [7], but further analysis is recommended for different cases.

These results can be visualized on the graph of **Fig. 3a**, where the joint shear strength index, τ, is plotted against the approximated normal compressive stress on

FIG. 2 Definition of the stresses in a shove test [8].

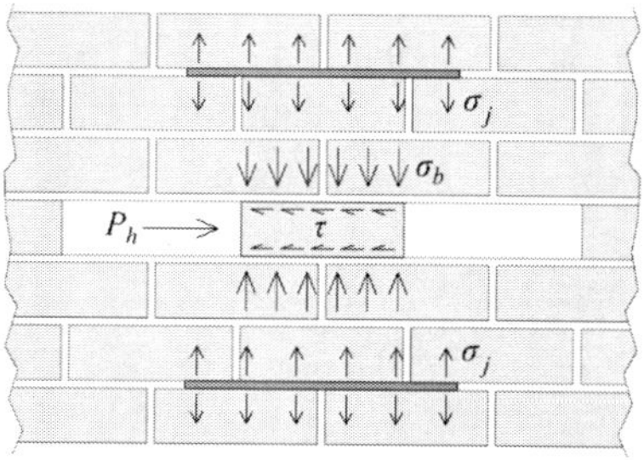

FIG. 3 Interpretation of the shove test: (a) according to "Method A" of ASTM C1531-16; (b) taking into account the peak strength due to cohesion.

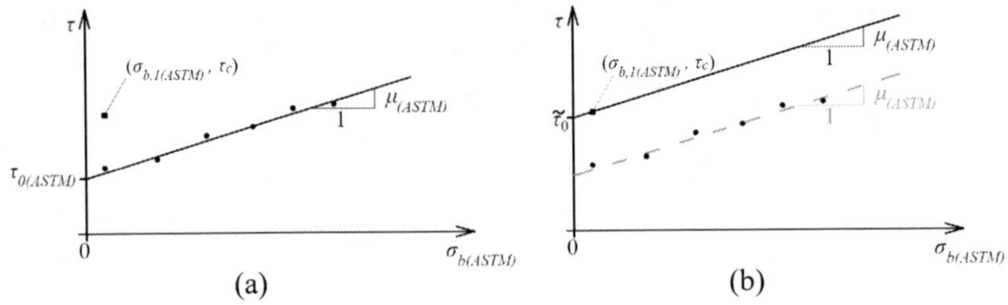

the test brick, $\sigma_{b(ASTM)}$. The relationship can be expressed by the frictional equation (Eq 4):

$$\tau = \tau_{0(ASTM)} + \mu_{(ASTM)} \cdot \sigma_{b(ASTM)} \tag{4}$$

The friction coefficient, $\mu_{(ASTM)}$, and the adhesion stress, $\tau_{0(ASTM)}$, can be taken as the slope and the y-intercept of the best-fitting line through the data points, respectively, as shown in Eq 5 and Eq 6:

$$\mu_{(ASTM)} = \frac{\sum_{i=1}^{N}\left[\left(\sigma_{b,i(ASTM)} - \bar{\sigma}_{b(ASTM)}\right)\left(\tau_i - \bar{\tau}\right)\right]}{\sum_{i=1}^{N}\left(\sigma_{b,i(ASTM)} - \bar{\sigma}_{b(ASTM)}\right)^2} \tag{5}$$

$$\tau_{0(ASTM)} = \bar{\tau} - \mu \cdot \bar{\sigma}_{b(ASTM)} \tag{6}$$

where a least-square linear regression has been performed, with sampling means $\bar{\sigma}_{b(ASTM)} = \left(\sum_{i=1}^{N} \sigma_{b,i(ASTM)}\right)/N$ and $\bar{\tau} = \left(\sum_{i=1}^{N} \tau_i\right)/N$, and N representing the number of levels of imposed flatjack pressure.

Issues with ASTM C1531-16 Procedure and Interpretation

NEGLECTING THE INITIAL PEAK STRENGTH DUE TO COHESION

As noted earlier, the purpose of the test is to estimate the Coulomb frictional parameters necessary to compute the shear strength of brick masonry in case of through-joints failure. Analysis of the bed-joint interface mechanical behavior shows that only the initial peak shear strength, immediately preceding the onset of sliding, is influenced by the material cohesion due to chemical bond. For this reason, discarding the first point associated with normal compressive stress $\sigma_{b,1(ASTM)}$ and shear stress τ_c, as implied by ASTM C1531-16 §9.1.2 [1], leads to an incorrect estimation of the cohesion.

An easy, yet misleading attempt to account for the first point could be: (i) computing the friction coefficient, $\mu_{(ASTM)}$, based on the pairs $\left(\sigma_{b,i(ASTM)}, \tau_i\right)$ measured during brick sliding per the current standard; then (ii) determining the cohesion as the y-intercept, $\tilde{\tau}_0$, of the best-fitting line translated through the point $(\sigma_{b,1(ASTM)}, \tau_c)$ at peak strength (**Fig. 3b**). This is indeed similar to the interpretation of Method B in ASTM C1531-16 [1]. However, this solution is incomplete and could lead to unconservative overestimations of the cohesion due to the factors discussed in the following paragraphs.

INACCURACY OF THE JACK-TO-BRICK MODIFICATION FACTOR

The definition of the jack-to-brick modification factor, j, according to Appendix X1 of ASTM C1531-16 [1], depends on the test configuration; the proposed value of 1.7 is unique for the specific case illustrated. As stated in the standard, "Further analysis would have to be conducted to determine the actual state of normal stress acting on the test unit for other geometries and test configurations" [1]. An incorrect value of j leads to an incorrect estimation of the friction coefficient, μ. Two rational methods are proposed in next sections to compute the stress on the test unit, σ_{bj}, due to the flatjack pressure.

NEGLECTING THE FAR-FIELD EFFECT

ASTM C1531-16 [1] does not take into account the fact that the vertical stress on the test brick is not zero when the flatjacks are unloaded. This initial stress is due to the far-field effect of the overburden stress acting on the surrounding masonry (including its self-weight), as confirmed by experimental and numerical studies [8,9]. **Fig. 4a** plots the residual shear strength index of a selected test unit in a double-leaf clay-brick masonry wall subjected to different overburden stresses, σ_{wall}. One can notice that, given the same flatjack pressure, the far-field stress affects the residual shear strength index of the tested unit. Furthermore, **Fig. 4b** plots the results of a calibrated

FIG. 4 Influence of the overburden load on the shear strength index: (a) residual shear strength from laboratory shove tests on a double-wythe clay-brick masonry wall [8]; (b) normal stress distribution due to far-field effect from numerical simulation of a shove test on a single-leaf calcium-silicate masonry wall [9]. Conversion: 1 MPa = 145 psi.

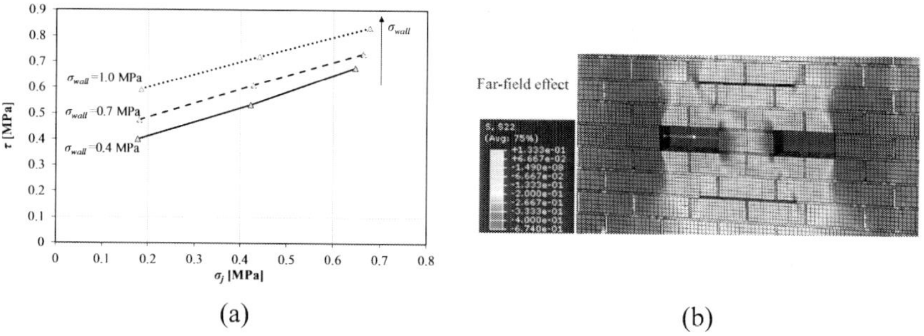

(a) (b)

numerical simulation of the shove test on a single-leaf calcium-silicate wall; the vertical compressive stress on the brick, σ_b, is non-zero even before loading the flatjacks.

Fig. 5 plots the time-history of the normal stress on the test unit recorded during the test numerical simulation. It is possible to observe that the far-field vertical stress, σ_{bff}, due to the overburden load, σ_{wall}, represents a non-negligible portion of the total compressive stress on the tested brick, σ_b. An incorrect estimation of the total compressive stress, σ_b, which does not consider the far-field component, σ_{bff}, leads to a wrong estimation of the cohesion, τ_0.

NEGLECTING THE DILATANCY EFFECT

Dilatancy is the volume change observed in granular materials when they are subjected to shear deformations. A detailed study on its effect on the shove test was conducted by Andreotti, Graziotti, and Magenes [9]. As will be discussed in more in detail later in this paper, the irregular sliding surface may be seen as a combination of asperities with different sizes [10]. Primary asperities, larger in size, are responsible for bed-joint dilation; the slope of primary asperities, ψ, usually is termed the dilatancy angle (Fig. 6). The friction coefficient, μ, is instead related to smaller secondary asperities across the inclined interfaces of the primary asperities.

The normal stress time-history of Fig. 5 shows that restraining the expansion of the specimen, which would be induced by dilatancy upon shearing, increases the normal compression on the tested brick by a value σ_{bdil}. This stress increase can be observed if the analysis is performed with the experimentally calibrated dilatancy angle, $\psi \neq 0$, although it is not visible if the same analysis is repeated with $\psi = 0$. This effect is more pronounced in close proximity to the initial peak shear strength,

FIG. 5 Contributions to the normal compressive stress on the test brick in a numerical simulation of a shove test: far-field effect, flatjack pressure, and dilatancy. Conversion: 1 MPa = 145 psi.

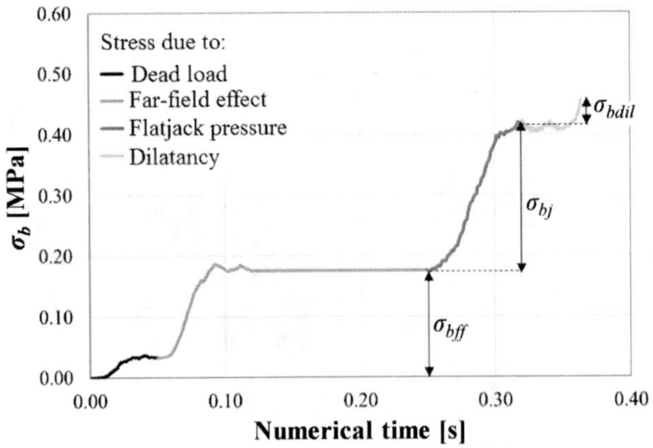

FIG. 6 Theoretical model of mortar joint dilation under shearing deformation [9].

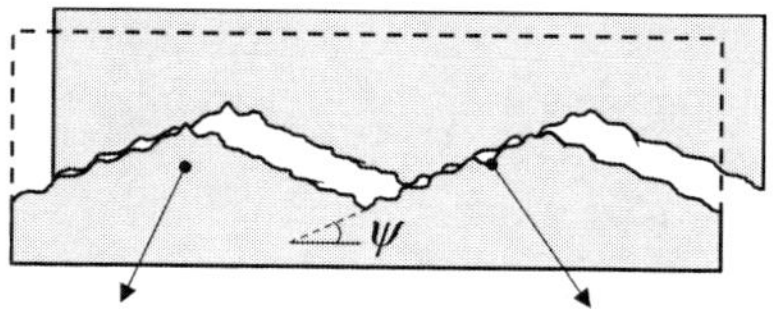

τ_c, when the bond between brick and mortar is breaking; it becomes less evident during further shoves of the tested unit. If the normal compressive stress associated with the peak shear strength, τ_c, is underestimated by neglecting the dilatancy contribution, σ_{bdil}, the cohesion could be unconservatively overestimated.

Proposed Amendments to ASTM C1531-16

In this section, a modified procedure of performing and interpreting the shove test is outlined that allows more accurate evaluation of the in situ effective compressive stress on the test unit. A more reliable assessment of the compressive stress on the test unit, in turn, will result in a more rational shear strength index evaluation. The relationship between the normal compressive stress in the test brick, σ_b, and the joint shear stress index, τ, is described by Coulomb's law (Eq 7):

$$\tau = \tau_0 + \mu \cdot \sigma_b \qquad (7)$$

The effective compressive stress on the test unit, σ_b, is decomposed into the three contributions illustrated in **Fig. 5** and introduced earlier (Eq 8):

$$\sigma_b = \sigma_{bj} + \sigma_{bff} + \sigma_{bdil} \qquad (8)$$

The stress contributions σ_{bj}, σ_{bff}, and σ_{bdil} will be discussed in detail in the following sections. It has been pointed out that dilatancy has a significant effect on the normal compressive stress especially when reaching the initial peak shear strength, τ_c, while its contribution decays as the test unit starts sliding or the normal compression increases, or both. However, the rate of decay is not clearly defined, and dilatancy may still have some influence especially during sliding right after the initial peak strength. For this reason, the proposed method retains the initial peak point, $(\sigma_{b,1}, \tau_c)$, but it discards the one associated with the residual shear strength measured during sliding under the same flatjack pressure $(\sigma_{b,1}, \tau_1)$.

FLATJACK PRESSURE CONTRIBUTION

Two approaches are presented herein to determine the contribution, σ_{bj}, of the flatjack pressure to the normal compressive stress on the test brick. The first one has a more rational mechanical background; it is based on the ratio of the elastic moduli of the masonry, measured before and after removal of the adjacent bricks, according to the double-flatjack method of ASTM C1197-14a [6]. The second approach is simpler and considers only the dimensions of the test brick and of the flatjacks.

Refined Approach: Elastic Moduli Ratio

This approach requires the double-flatjack test of ASTM C1197-14a [6] to be repeated twice: first, on the intact wall before removing the bricks adjacent to the test unit; second, after removal of the adjacent units. The location of the flatjacks and of the vertical displacement transducers must be the same in both configurations. However, it is generally recommended to remove the vertical displacement transducers during the cutting operations and to reattach them in the original positions. The first double-flatjack test provides the conventional elastic modulus of the masonry, E. Repeating the test after removal of the two adjacent units results in a fictitious elastic modulus, E^*, which is affected not only by the intrinsic material properties but also by the geometry of the shove-test configuration. It is assumed that the intrinsic material properties remain the same in both configurations. For this scope, the double-flatjack test of ASTM C1197-14a [6] should be conducted without exceeding the elastic range of response of the masonry.

The stress flow within the wall due to flatjack pressure, after removal of the two bricks, is qualitatively depicted in **Fig. 7a**. Depending on the geometry of the bricks and on the number of wythes, the normal compressive stress on the test unit, σ_{bj}, can be larger or smaller than the stress on the masonry in contact with the flatjacks, σ_j. The axial compressive strain, ε_m, of the masonry in close proximity of the test brick in the shove-test configuration can be expressed as in Eq 9:

$$\varepsilon_m = \frac{\sigma_{bj}}{E} \tag{9}$$

FIG. 7 Flatjack pressure contribution to the normal stress on the test brick: (a) qualitative stress flow in the shove-test configuration [8]; (b) graphical representation of friction coefficient and residual shear strength index.

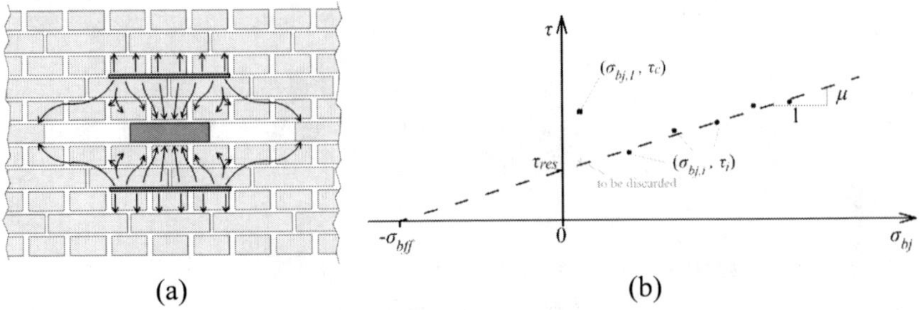

However, the same axial compressive strain can be interpreted as the result of the stress on the masonry in contact with the flatjacks applied over the wall portion characterized by a fictitious elastic modulus, E^* (Eq 10):

$$\varepsilon_m = \frac{\sigma_j}{E^*} \tag{10}$$

Combining Eq 9 and Eq 10 leads to a relationship between the normal compressive stress on the test unit, σ_{bj}, and the stress on the masonry in contact with the flatjacks, σ_j, as shown in Eq 11:

$$\sigma_{bj} = k_{bj}\sigma_j \tag{11}$$

where the modification factor is given as (Eq 12):

$$k_{bj} = \frac{E}{E^*} \tag{12}$$

Simplified Approach: Flatjack-to-Brick Length Ratio

Experimental results from single- and double-wythe walls, tested in the configuration illustrated in **Fig. 1**, show that the modification factor, k_{bj}, could be approximatively estimated considering only geometric properties of the test setup (Eq 13):

$$k_{bj} = \frac{A_{je}}{N_w \cdot A_b} \tag{13}$$

where A_{je} is the effective contact area between the flatjacks and only the masonry wythe containing the test brick; A_b is the area of the horizontal cross-section of the test brick; $N_w = 1$ for single-wythe walls; and $N_w = 1.5$ for double-wythe walls. Data are currently not available for walls with more than two wythes. Usually, A_{je} represents about 90 % to 95 % of the area of the flatjack in contact with the masonry wythe, which contains the test brick, due to edge effects at the flatjack welds; it can be estimated in situ once the flatjacks are loaded. In the case of double-leaf walls, the portion of the flatjack area extending into the back wythe, if any, should not be considered to determine A_{je}.

When this procedure is adopted, evaluating the fictitious elastic modulus, E^*, is not necessary, and only one double-flatjack test on the intact masonry wall could be performed to determine its conventional elastic modulus, E.

DETERMINATION OF THE FRICTION COEFFICIENT

Once the normal stresses on the test brick, $\sigma_{bj,i}$, have been calculated for all levels of flatjack pressure, the friction coefficient, μ, can be evaluated as the slope of the best-fitting line through all test points $(\sigma_{bj,i}, \tau_i)$ with $i = 2, \ldots N$ (**Fig. 7b**). If a least-square linear regression is performed, the friction coefficient is given by Eq 14:

$$\mu = \frac{\sum_{i=2}^{N}\left[\left(\sigma_{bj,i} - \bar{\sigma}_{bj}\right)\left(\tau_i - \bar{\tau}\right)\right]}{\sum_{i=2}^{N}\left(\sigma_{bj,i} - \bar{\sigma}_{bj}\right)^2} \tag{14}$$

where the sampling means are $\bar{\sigma}_{bj} = \left(\sum_{i=2}^{N} \sigma_{bj,i}\right)/(N-1)$ and $\bar{\tau} = \left(\sum_{i=2}^{N} \tau_i\right)/(N-1)$. As stated earlier, the point corresponding to sliding under the initial level of compression is discarded because the normal stress on the brick would be affected by dilatancy to some degree. The point corresponding to the initial peak shear strength, $(\sigma_{bj,1}, \tau_c)$, is not considered for the determination of μ as it is dominated by cohesion effects.

Despite its similarity to the modification factor, j, defined in Appendix X1 of ASTM C1531-16 [1], the proposed factor, k_{bj}, has an explicit physical meaning. While j accounts for the difference between σ_{bj} and σ_j on a statistical basis, possibly including the far-field effects of the overburden stress, k_{bj} addresses the stress flow due to the flatjacks only. The far-field effect is treated separately and provides an additional normal stress contribution, as will be discussed in the following section.

FAR-FIELD EFFECT CONTRIBUTION

The best-fitting line obtained after correction of the normal compressive stress with k_{bj} only (**Fig. 7b**) intercepts the y-axis at a non-zero residual shear strength index (Eq 15):

$$\tau_{res} = \bar{\tau} - \mu \cdot \bar{\sigma}_{bj} \tag{15}$$

However, since this line represents a frictional strength after bond breaking, it would be expected to provide zero residual shear strength for $\sigma_b \approx 0$, as obtained from laboratory direct shear tests [7,11,12]; for instance, **Fig. 8a** shows the results of a triplet test conducted by Graziotti et al. [12]. This highlights the fact that σ_{bj} does not constitute the actual normal compressive stress on the brick-mortar interface.

FIG. 8 Far-field contribution to the normal stress on the test brick: (a) residual shear strength from a laboratory triplet test on a clay-brick masonry specimen [12]; (b) qualitative stress flow in the shove-test configuration [8]. Conversion: 1 MPa = 145 psi.

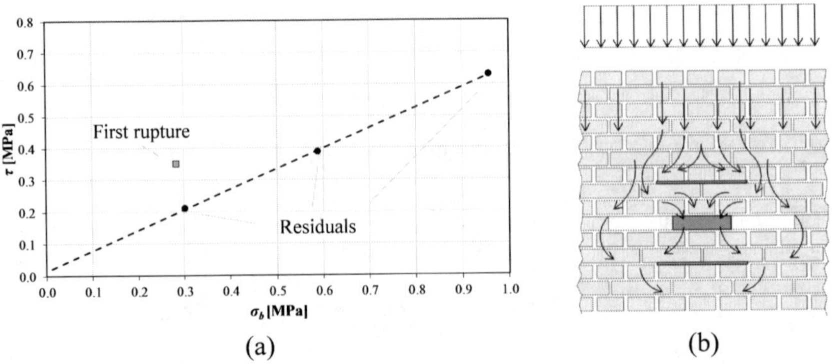

An additional contribution to the normal compressive stress on the test unit, σ_{bff}, comes from the far-field effect, as qualitatively illustrated in **Fig. 8b**. The magnitude of this contribution depends on the self-weight and on the existing overburden stress prior to the flatjack insertion and brick removal; the geometry of bricks and slots and the number of wythes also influences the far-field stress component. This effect is particularly evident on heavily loaded, multi-wythe walls but can be noticed also in single-wythe walls subjected to compressive stresses as small as approximately 0.1 to 0.2 MPa (15 to 30 psi).

The far-field effect on the normal stress on the test unit, σ_{bff}, can be accounted for in a rather simple way once the normal stress due to the flatjacks, $\sigma_{bj,i}$, has been determined for all test points. Imposing the shear strength index after bond rupture to be zero if $\sigma_b \approx 0$ is equivalent to translating the points and the best-fitting dashed line of **Fig. 7b** to the right, until the line passes through the axis origin. From geometry, one can derive Eq 16:

$$\sigma_{bff} = \frac{\tau_{res}}{\mu} \tag{16}$$

which is the same for any level of flatjack pressure and all test points ($i = 1, \ldots N$).

It is worth noting that this additional normal stress contribution does not affect the value of the friction coefficient, μ. In fact, adding σ_{bff} to all normal compressive stresses shifts all data points of **Fig. 7b** to the right by the same amount, leaving the slope of the best-fitting line unchanged.

Alternative solutions to addressing the aforementioned issues may also include variations in the execution of the test. For example, longer horizontal cuts may help reduce the influence of far-field stresses on the test unit normal compression at the expense of increased invasiveness and operational difficulties. In this case, the flatjacks may need to be shorter than the slots to avoid compromising the calculation of σ_{bj}. However, additional laboratory testing is necessary to determine the optimum geometrical configuration for this solution.

DILATANCY CONTRIBUTION

The onset of sliding in a shove test requires de-cohesion of the test brick. This process consists of a loss of chemical bond within the mortar and over the brick-mortar interface, with formation and propagation of cracks that define a sliding surface. The mechanical behavior of this surface can be described as a combination of friction and normal deformation, which are governed by roughness and inclination of the cracking surface, respectively. Depending on the level of compression relative to the mortar strength, the onset of sliding can be accompanied by normal contraction or dilation: When the compression level is close to the mortar compressive strength, joints tend to contract, otherwise they dilate. Generally, the normal compressive stresses found in situ are relatively far from the mortar strength, making dilation more likely than contraction.

According to Stupkiewicz and Mróz [10], the irregular sliding surface may be seen as a combination of asperities with different sizes (**Fig. 6**). Primary asperities are larger and responsible for bed-joint dilation; the slope of primary asperities usually is termed dilatancy. Smaller secondary asperities govern friction across the inclined interfaces of the primary asperities. Although coupling exists between the effects of the two type of asperities, this simplified assumption allows separation of the dominant mechanisms. Experimental tests [13] show the dilation behavior varies with the magnitude of sliding shear displacement and normal compression. Larger dilatancy angles can be observed upon first-cracking and during the initial stages of sliding, especially if the normal compressive stress is low. When the relative shear displacement approaches the length of the primary asperities, dilatancy angle and joint dilation start to decrease due to crushing of these asperities (**Fig. 9**), particularly under high normal compression.

The sliding brick of a shove test is confined top and bottom by the surrounding masonry and is compressed by the flatjacks and by the overburden loads; dilation of the two mortar joints is partially restrained in this configuration. The onset of sliding results then in additional normal stress on the tested brick, σ_{bdil}, to overcome the clamping effect of the surrounding material. **Fig. 10** illustrates the results of calibrated numerical simulations of shove tests on two single-leaf calcium-silicate walls [9]. In particular, in **Fig. 10a**, the average compressive stress on the test brick, σ_b, is plotted against the numerical time until reaching the peak shear strength. Both walls have an initial elastic modulus equal to 4,182 MPa (607 ksi). Dilatancy angles ψ of 10° and 15°, experimentally recorded at the onset of sliding, characterize Tests #1 and #2, respectively. The shove phase is characterized by an increment in the average compressive stress due to dilatancy in both cases, with higher values for larger ψ; the overstress due to dilatancy, σ_{bdil}, represents 10 % and 33 % of the total compressive stress on the brick at peak shear strength, $\sigma_{b,1}$, for Tests #1 and #2, respectively.

Fig. 10b illustrates the shear strength index, τ, and the vertical dilation, *dy*, against the sliding displacement of the brick during the shove phase. Comparing **Fig. 10a** and **b** shows that a higher increment of normal stress on the brick is associated with a larger vertical dilation when the shear strength reaches its peak value, τ_c. Based on these numerical results, two methods are proposed in the following paragraphs to account for the effect of dilatancy on the normal compressive stress of the test unit. Measuring the dilation to estimate the stress, σ_{bdil}, may require high-precision in situ operations that is not always feasible; for this reason, one of

FIG. 9 Primary asperities and dilatancy at different stages of sliding [9].

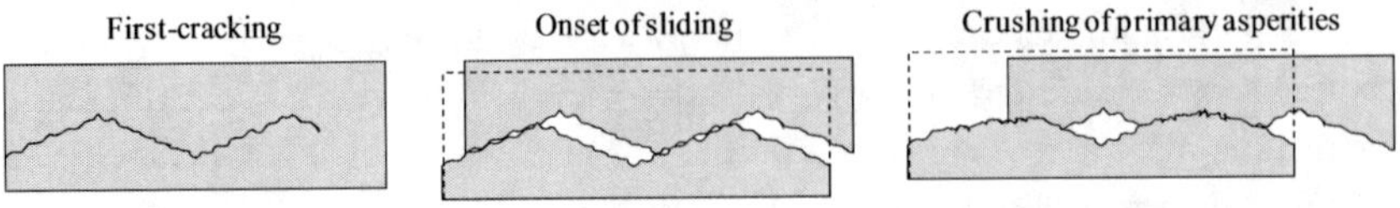

FIG. 10 Dilatancy contribution to the normal stress on the test brick from numerical simulation of a shove test [9]: (a) normal compressive stress history; (b) shear stress and dilation during the shove phase. Conversions: 1 MPa = 145 psi, 1 mm = 0.0394 in.

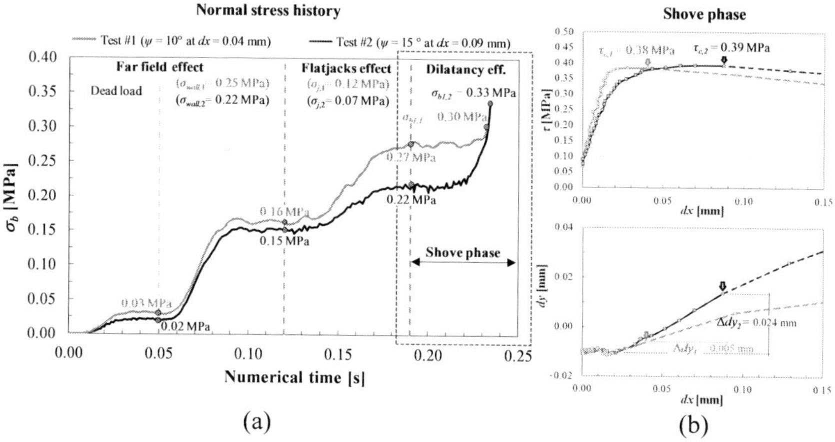

the two methods allows for a lump-estimate of the stress increment without direct measurements.

Since both the dilatancy angle and dilation reduce as sliding displacement and normal compression increase, only the first data point, $(\sigma_{b,1}, \tau_c)$, can be significantly influenced by this effect. For this reason, it is assumed that $\sigma_{bdil,1} \neq 0$ when associated with τ_c, while $\sigma_{bdil,i} = 0$ for all other test points with $i = 2, \ldots N$. The point associated with the residual shear strength under the first level of flatjack pressure should be discarded as a precaution; in fact, the dilatancy effect is characterized by large uncertainties in the immediate post-peak phase.

Refined Approach: Vertical Dilation Measurement

The application of this method requires measurement of the vertical dilation of the two bed joints, using one or two vertical displacement transducers attached to the masonry above and below the test brick between the flatjacks (**Fig. 11**). The displacement transducers can be installed at the same locations chosen for the double-flatjack test. This approach takes advantage of the numerical observation that the vertical overstress due to dilatancy, σ_{bdil}, is proportional to the vertical dilation, dy.

The masonry surrounding the test brick is idealized with two linear springs, one above and one below the brick, subjected to additional compression by the dilatancy effect and restrained against vertical displacements far away from the test location. Each spring is assigned the masonry elastic modulus, E; a cross-sectional area equal to the area of the test brick, A_b; and a length, $k_L \cdot L_b$, proportional to the length of the brick L_b. The length factor, k_L, accounts for the simplified geometric

FIG. 11 Modified shove test setup.

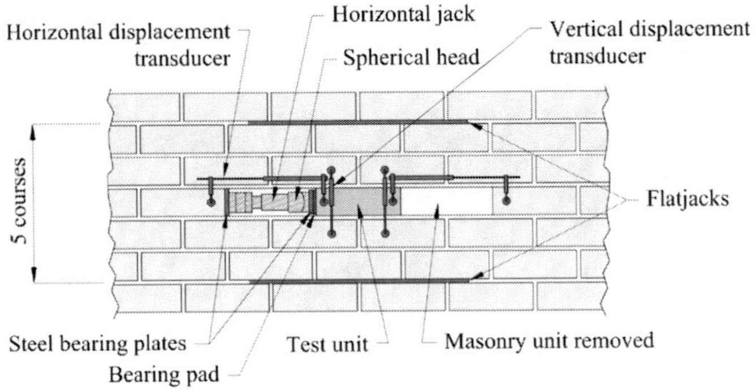

assumptions on the spring cross section and is physically related to the attenuation length of local effects. The contribution of dilatancy to the normal stress can be computed as Eq 17:

$$\sigma_{bdil,i} = \begin{cases} \dfrac{\Delta dy \cdot E}{2 \cdot k_L \cdot L_b} & \text{for } i = 1 \\ 0 & \text{for } i = 2, \ldots N \end{cases} \tag{17}$$

where Δdy represents the dilation, which is the displacement measured by the vertical transducers between the beginning of the shove phase and the instant when the peak shear strength, τ_c, is reached. The length factor, k_L, has been evaluated equal to 2.5 considering the single-wythe calcium-silicate masonry walls analyzed numerically by Andreotti, Graziotti, and Magenes [9] and the double-wythe clay-brick walls described in Rossi, Graziotti, and Magenes [8].

Simplified Approach: Lumped Overstress Calculation

This method simply takes the dilatancy contribution to the normal compressive stress, σ_{bdil}, as a percentage of the total vertical stress on the test unit without direct measurement of the vertical dilatation. The numerical results of **Fig. 10** show the dilatancy contribution is between 10 % and 33 % of the total compressive stress on the brick at peak shear strength or between 10 % and 50 % of the stress due to flatjack pressure and far-field effects. It is then possible to estimate the contribution of dilatancy to the normal compressive stress as Eq 18:

$$\sigma_{bdil,i} = \begin{cases} k_{dil} \cdot \left(\sigma_{bj,1} + \sigma_{bff}\right) & \text{for } i = 1 \\ 0 & \text{for } i = 2, \ldots N \end{cases} \tag{18}$$

where factor k_{dil} theoretically is a function of the dilatancy angle, ψ. Numerical results on single-wythe calcium-silicate masonry walls by Andreotti, Graziotti, and Magenes [9] and experimental evidence on double-wythe clay-brick walls by Rossi,

Graziotti, and Magenes [8] suggest that k_{dil} can be set between 0.1 (lower values of ψ) and 0.5 (higher values of ψ).

DETERMINATION OF THE COHESION

The total normal stresses on the test brick can be calculated for all test points ($i = 1, \ldots N$), including the first one corresponding to the peak shear strength index, as Eq 19:

$$\sigma_{b,i} = \sigma_{bj,i} + \sigma_{bff} + \sigma_{bdil,i} \tag{19}$$

If a straight line with slope equal to μ is drawn on the (σ_b, τ) plane through the point $(\sigma_{b,1}, \tau_c)$, its y-intercept represents the cohesion, τ_0, as shown in **Fig. 12** (Eq 20):

$$\tau_0 = \tau_c - \mu \cdot \sigma_{b,1} \tag{20}$$

Proposed Testing Procedure and Interpretation

STEP-BY-STEP TESTING PROCEDURE

If the simplified approaches introduced earlier are employed for the determination of both k_{bj} and σ_{bdil}, the standard testing procedure termed Method A in ASTM C1531-16 [1] can be followed as summarized at the beginning of this paper; in this case, only the data interpretation differs from the current ASTM standard. Instead, additional steps are required if the refined approaches are chosen. The complete procedure is outlined here:

1. Identify the brick unit to be tested, which must be in the stretcher orientation; locate the test site far enough from wall openings or ends so that the loading jack bears against enough masonry to resist forces generated during loading of the test unit.
2. Prepare two slots in the masonry bed joints, two courses above and two courses below the test unit; insert the flatjacks in the two slots, ensuring they are parallel to each other and vertically aligned with the test brick, according to ASTM C1197-14a [6].

FIG. 12 Graphical representation of the corrected shear strength Coulomb-type relationship.

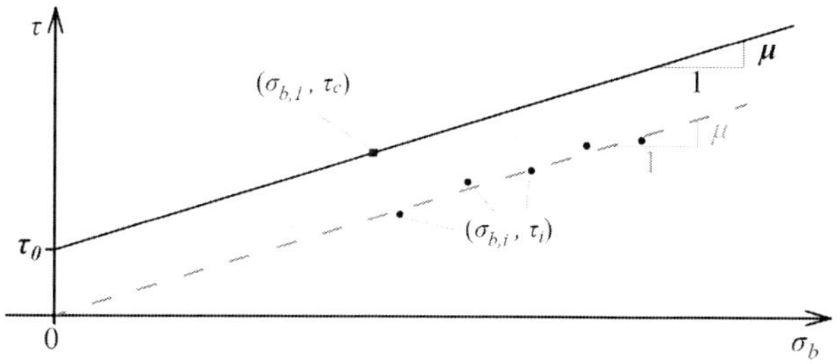

3. Perform a double-flatjack test on the intact wall before removing the bricks adjacent to the test unit, according to ASTM C1197-14a [6], to determine the masonry elastic modulus, E.
4. Remove the two bricks adjacent to the test unit; for multiple-leaf walls, also remove the vertical mortar joint on the back side of the test brick.
5. If the refined approach is chosen to determine k_{bj}, perform a double-flatjack test on the wall in the shove-test configuration, according to ASTM C1197-14a [6], to determine the fictitious elastic modulus, E^*.
6. Instrument the test unit with one or two horizontal displacement transducers; place the loading jack, spherical seat, and bearing plates with bearing pads into the space next to the test unit; shim the jack to provide a centered horizontal force on the test unit.
7. If the refined approach is chosen to determine σ_{bdil}, install one or two vertical displacement transducers, connecting them to the masonry above and below the test unit between the flatjacks (**Fig. 11**).
8. Set the flatjack pressure equal or less than 0.07 MPa (10 psi) and hold it constant; increase pressure in the horizontal loading jack gradually, recording displacement of the unit at small increments of horizontal load; when the bond between the unit and mortar joints is broken, the unit will begin to displace continuously under constant horizontal load; the peak load recorded right before sliding represents the maximum horizontal load for this first level of normal compressive stress.
9. Increase the pressure in the flatjacks to the next desired value and repeat the process of horizontal loading until the maximum horizontal load for this level of normal compressive stress is reached; repeat the sequence in this manner to determine the bed-joint residual shear stress at various levels of normal compressive stress; be sure not to exceed the elastic branch of the compressive force-displacement envelope determined with the double-flatjack test of Step (3).
10. To investigate the effect of shear force reversal on the in situ shear strength, if desired, the horizontal jack may be transferred to the opening on the opposite side of the test unit, the displacement instrumentation reversed, and the test sequence repeated in the opposite direction.
11. Release pressure in the horizontal jack after the final displacement measurement has been taken and remove the loading jack; release pressure in the flatjacks and remove the flatjacks and displacement measurement devices.

DATA INTERPRETATION

A step-by-step procedure is proposed to interpret the results of the shove test, incorporating the contributions to the normal compressive stress analyzed earlier in detail:

1. For the first level of normal compression, determine the mortar joint shear strength index, τ_c, corresponding to the peak force immediately preceding the onset of sliding, P_{hc}, using Eq 1. Discard the point corresponding to the residual horizontal force under the first level of normal compression. Apply Eq 2 to calculate the shear strength index, τ_i, corresponding to the residual horizontal force, $P_{h,i}$, for the other levels of normal compression, $i = 2, \ldots N$.

2. Record the vertical stress applied by the flatjacks to the masonry in contact with them, $\sigma_{j,i}$, at all i^{th} levels of compression, in accordance with the ASTM C1197-14a [6].
3. Determine the modification factor, k_{bj}, using Eq 12 for the accurate approach or Eq 13 for the simplified one. Then calculate the normal compressive stress on the test unit due to flatjack pressure, $\sigma_{bj,i}$, applying Eq 11 for $i = 1, \ldots N$.
4. Evaluate the friction coefficient, μ, as the slope of the best-fitting line through all test points $(\sigma_{bj,i}, \tau_i)$ with $i = 2, \ldots N$. Eq 14 can be used if a least-square linear regression approach is chosen. The point corresponding to the initial peak shear strength, $(\sigma_{bj,1}, \tau_c)$, is not considered for this purpose.
5. Calculate the residual shear strength index, τ_{res}, as the y-intercept of the best-fitting line through all test points $(\sigma_{bj,i}, \tau_i)$ with $i = 2, \ldots N$; Eq 15 can be used if a least-square linear regression approach is chosen. Then determine far-field normal stress contribution, σ_{bff}, applying Eq 16; this value is the same for all data points, $i = 1, \ldots N$.
6. Determine the contribution of dilatancy to the normal stress, applying Eq 17 for the accurate approach or Eq 18 for the simplified one. Then calculate the total normal compressive stresses on the test unit using Eq 19 for $i = 1, \ldots N$.
7. Apply Eq 20 to evaluate the cohesion, τ_0, as the y-intercept of the line with slope equal to μ, drawn on the (σ_b, τ) plane through the point $(\sigma_{b,1}, \tau_c)$.

Conclusions

This paper discussed a modified interpretation of the shove test, an in situ method to evaluate the shear strength of masonry walls in case of through-joints failure. This test aims at estimating the friction coefficient and cohesion of a Coulomb-type relationship between the shear strength index and the normal compressive stress on the masonry. Two horizontal flatjacks impose a controlled vertical compression on a portion of the masonry wall, while a brick unit located in the masonry portion between the flatjacks is pushed horizontally.

The interpretation of the test outcomes is affected by difficulties in determining the actual normal compressive stress acting on the test brick. The current ASTM C1531 standard suggests estimating this stress as a multiple of the one applied by the flatjacks to the masonry in contact with them and provides a modification factor that is valid only for a particular geometric configuration. However, this approach does not consider the actual stress flow due to the applied flatjack pressure and to the overburden loads acting on the wall and the effect of dilatancy at the onset of sliding. Failing to properly estimate the normal compression on the test unit may result in significant errors in the frictional strength parameters.

An alternative interpretation was proposed for calculating more accurately the effective normal stress on the test brick as the sum of three contributions based on experimental and numerical results. The first contribution comes from the applied flatjack pressure, considering the disturbance on the stress flow due to removal of the bricks adjacent to the test unit. The second component is caused by the far-field effect of the overburden loads. The last contribution is due to the restrained expansion of the bed joints due to dilatancy at the onset of sliding. The proposed amendments aim at improving the accuracy of the method despite the intrinsic variability affecting the precision of the test results. It appears that these changes are beneficial also in terms of precision, but more tests are needed to confirm this.

Refined and simplified approaches were developed to estimate the first and the third contributions. The simplified procedures do not require any additional instrumentation or test steps compared to the current ASTM standard. When higher accuracy is desired, the refined methods can be applied; however, these require performing two double-flatjack tests and installing vertical displacement transducers across the test brick.

Further laboratory tests, in situ campaigns, and numerical simulations can be performed to extend the validity of the proposed method to different masonry materials and geometric configurations.

ACKNOWLEDGMENTS

The work presented in this paper was partly supported by the financial contribution of the Italian Department of Civil Protection within the framework "RELUIS-DPC," which is greatly acknowledged by the authors. The performed laboratory tests were part of the project "Study of the Vulnerability of Masonry Buildings in Groningen," carried out at EUCENTRE within the framework of the research program for hazard and risk of induced seismicity in Groningen, sponsored by the Nederlandse Aardolie Maatschappij, BV (NAM). The authors would like to thank all the partners involved in this project: the DICAr laboratory of the University of Pavia and the EUCENTRE laboratory that performed the tests, together with NAM, Arup, and TU Delft. Special thanks are due to C. Lai for his suggestions. The authors would like to express their gratitude also to S. Girello, F. Dacarro, R. Pistore, V. Bonura, S. Peloso, and P&P Consulting Engineers Group for the execution of the tests at DICAr laboratory.

References

[1] ASTM C1531-16, *Standard Test Methods for In Situ Measurement of Masonry Mortar Joint Shear Strength Index*, ASTM International, West Conshohocken, PA, 2016, http://dx.doi.org/10.1520/C1531-16

[2] EN 1052-3:2002, *Methods of Test for Masonry—Part 3: Determination of Initial Shear Strength*, European Committee for Standardization (CEN/TC 125), Brussels, Belgium, 2002, www.cen.eu

[3] EN 1052-3:2002/A1:2007, *Amendment A1 to European Standard EN 1052-3:2002*, European Committee for Standardization (CEN/TC 125), Brussels, Belgium, 2007, www.cen.eu

[4] Calvi, G. M., Kingsley, G. R., and Magenes, G., "Testing of Masonry Structures for Seismic Assessment," *Earthq. Spectra*, Vol. 12, No. 1, 1996, pp. 145–162, http://dx.doi.org/10.1193/1.1585872

[5] Magenes, G. and Calvi, G. M., "In-Plane Seismic Response of Brick Masonry Walls," *Earthq. Eng. Struct. D.*, Vol. 26, No. 11, 1997, pp. 1091–112, https://dx.doi.org/10.1002/(SICI)1096-9845(199711)26:11<1091::AID-EQE693>3.0.CO;2-6

[6] ASTM C1197-14a, *Standard Test Method for In Situ Measurement of Masonry Deformability Properties Using the Flatjack Method*, ASTM International, West Conshohocken, PA, 2014, http://dx.doi.org/10.1520/C1197-14A

[7] Atkinson, R. H., Kingsley, G. R., Saeb, S., Amadei, B., and Sture, S., "A Laboratory and In Situ Study of the Shear Strength of Masonry Bed Joints," *Proceedings of the Eighth International Brick and Block Masonry Conference*, Trinity College, Dublin, Republic of Ireland, September 19–21, 1988, J. W. de Courcy, Ed., Elsevier, London, pp. 261–271.

[8] Rossi, A., Graziotti, F., and Magenes, G., "A Proposal for the Interpretation of the In Situ Shear Strength Index Test for Brick Masonry," *Proceedings of the Sixteenth Conference of the Italian National Association of Earthquake Engineering*, University of L'Aquila, L'Aquila, Italy, September 13–17, 2015, ANIDIS, Rome, Paper No. 2149.

[9] Andreotti, G., Graziotti, F., and Magenes, G., "Detailed Micro-Modelling of the Direct Shear Tests of Brick Masonry Specimens: The Role of Dilatancy," *Eng. Struct.*, Vol. 168, 2018, pp. 929–949.

[10] Stupkiewicz, S. and Mróz, Z., "Modelling of Friction and Dilatancy Effects at Brittle Interfaces for Monotonic and Cyclic Loading," *J. Theor. App. Mech.*, Vol. 39, No. 3, 2001, pp. 707–739.

[11] Atkinson, R. H., Amadei, B. P., Saeb, S., and Sture, S., "Response of Masonry Bed Joints in Direct Shear," *J. Struct. Eng.-ASCE.*, Vol. 115, No. 9, 1989, pp. 2276–2296, http://dx.doi.org/10.1061/(ASCE)0733-9445(1989)115:9(2276)

[12] Graziotti, F., Tomassetti, U., Rossi, A., Marchesi, B., Kallioras, S., Mandirola, M., Fragomeli, A., et al., *Shaking Table Tests on a Full-Scale Clay-Brick Masonry House Representative of the Groningen Building Stock and Related Characterization Tests*, Report EUC128/2016U, EUCENTRE, Pavia, Italy, 2016.

[13] Van Der Pluijm, R., "Out-of-Plane Bending of Masonry: Behavior and Strength," Ph.D. thesis, Eindhoven University of Technology, Eindhoven, The Netherlands, 1999.

STP 1612, 2018 / available online at www.astm.org / doi: 10.1520/STP161220170150

Muhammad Romeo N. Khan,[1] Fahim Ahmed,[1]
and Raquib Ahsan[1]

Ambient Vibrations of Unreinforced Masonry Buildings in Dhaka City

Citation
Khan, M. R. N., Ahmed, F., and Ahsan, R., "Ambient Vibrations of Unreinforced Masonry Buildings in Dhaka City," *Masonry 2018, ASTM STP1612*, N. V. Krogstad and W. M. McGinley, Eds., ASTM International, West Conshohocken, PA, 2018, pp. 34–55, http://dx.doi.org/10.1520/STP161220170150[2]

ABSTRACT
The main objective of this research was to identify the predominant or natural periods of targeted unreinforced masonry (URM) buildings by performing ambient vibration tests to compare the obtained structural predominant period with the structural period calculated by the formula given in the Bangladesh National Building Codes. The study also found a relationship between the structural period and different structural parameters of URM buildings. This paper presents an evaluation of ambient vibrations and identifies the predominant frequencies of 17 URM buildings. The buildings range from a single story to five stories and have regular and irregular plans and elevations. The buildings are located in Dhaka, the capital city of Bangladesh. These masonry buildings were constructed from 1960 to 1990 without any seismic guidelines. An attempt was made to correlate structural periods with heights and with total length of structure walls using MS Excel plots and a Statistical Package for the Social Sciences model. In order to find a realistic solution, multivariate regression analysis was selected for model development where the structural period was the dependable variable. The correlations through regressions were successfully identified.

Manuscript received September 10, 2017; accepted for publication February 28, 2018.
[1]Dept. of Civil Engineering, Bangladesh University of Engineering and Technology, Dhaka-1000, Bangladesh
M. R. N. K. https://orcid.org/0000-0003-4806-2035, F. A. http://orcid.org/0000-0002-3594-4525,
R. A. http://orcid.org/0000-0003-1063-0543
[2]ASTM International Symposium on *Masonry 2018* on June 26, 2018 in San Diego, CA, USA.

Keywords
unreinforced masonry (URM), ambient vibration test (AVT), Fourier spectra, modal analysis, predominant frequency, structural period

Introduction

In the event of an earthquake, apart from the existing gravity loads, horizontal loads are imposed on buildings. Earthquakes generate sudden impulse-type loads into buildings, setting them in vibratory motions. For unreinforced masonry (URM) buildings, the walls carry loads and provide the capacity to resist the demand created by these motions. URM buildings are vulnerable due to their poor performance under seismic actions. However, an understanding of the dynamic behavior of URM buildings when subjected to earthquake excitation is of major concern in many seismically threatened countries such as Bangladesh.

Dynamic characteristics have a key role in the seismic behavior and vulnerability of structures. In particular, it is important to know fundamental periods of vibration, both in the design of new buildings and in the assessment of existing ones. A building's seismic response is important. Monitoring of structural responses from ambient vibration is preferred because dynamic properties can be identified by analyzing ambient responses of the buildings. The ambient vibration test (AVT) is simple, inexpensive, and does not require any heavy excitation devices such as forced-vibration tests.

There are several empirical formulas for calculating predominant or natural periods of structures in the codes of various countries. But these formulas are based largely on observations of the dynamic response of multistoried steel and reinforced concrete (RC) structures. Masonry structures are included in the "other" structures category in Bangladesh National Building Code 1993 [1]. To the authors' best knowledge, no such structures were found that could be studied to determine a specific formula for the fundamental period of URM structures. Therefore, this paper attempts to ascertain a specific relationship for calculating the predominant or natural period of URM structures within their structural parameters.

Ambient vibration is excitation from environmental elements. It can be produced from cultural noises, natural sources, or micro-seismic tremors (or a combination thereof). Unlike forced vibration testing, the forces applied to the structure in ambient vibration testing, which is a nondestructive technique, are not controlled. The structure is assumed to be excited by wind, traffic, and human activity. The measurements, typically accelerations, are taken for a long duration to ensure that all the modes of interest are sufficiently excited. An AVT describes the linear behavior of structures because the amplitudes of the vibration are small [2]. The purpose of performing AVTs in general is to find out the dynamic characteristics and damping estimates of a structure. The most frequent use today is to identify natural frequencies and corresponding mode shapes of structures. Consequently, AVTs are being extensively used for structural health monitoring (SHM) of civil engineering structures. The identification of dynamic characteristics is a necessary and important task in the course of seismic design of civil engineering structures [3]. These tests aid in determining the dynamic behavior of structures under low vibrations.

Ambient vibration measurements by different sensing equipment reveal the actual dynamic behavior of structures under small vibrations. This behavior depends on structural properties such as geometry, material properties, and so on. It is also affected by environmental parameters such as temperature, soil-structure interaction, the participation of nonstructural elements, and so on. Clinton et al. [4] revealed that the environmental parameters changed ±4 % in vibration frequency at the Millikan Library in California. In 1936, Carder outlined the beginnings of ambient and forced vibration tests of structures by the U.S. Coast and Geodetic Survey for determining the fundamental periods of vibration for some high-rise buildings [5]. Crawford and Ward [6,7] revitalized the significance of this method showing that it can be used to identify the lowest frequencies and modal shapes of complete structures. Klopp and Griffiths [8] instrumented 14 URM buildings to assess the applicability of the simplified design formulas in the Australian earthquake codes, AS 2121, and another in the draft of AS 1170.4 for low-rise-URM buildings based mostly on observations of the dynamic response of multistoried steel and RC structures. Balkaya and Kalkan [9] analyzed 80 different shear wall-dominated RC buildings by using a three-dimensional finite element method and proposed a set of new empirical equations for estimating fundamental periods.

A series of AVTs were conducted by Turek, Thibert, Ventura, and Kuan [10] on three low-rise URM buildings located in the Chinatown district, Vancouver, Canada, to accumulate primary information on the dynamic properties of the buildings. Gentile and Saisi [11] conducted AVTs to assess the structural conditions of a seventeenth-century masonry bell tower (about 74 m high). They found a good match between the theoretical and experimental modal parameters for relatively low stiffness ratios in the most damaged regions of the tower. Gilles and McClure [12] researched expanding the global period database and created a database specific to Montreal, Canada.

Hancilar, Durukal, and Erdik [13] assessed the earthquake vulnerability of a four-story URM historical building (1869) in Istanbul by the Schmidt hammer and AVT. They proposed reinforced cement jacketing of the main load-carrying walls and application of fiber-reinforced plastic (FRP) bands to the secondary walls for seismic rehabilitation of the building. Crowley and Pinho [14] analyzed the seismic design codes around the world to estimate the period of vibration of RC buildings for use in both linear static and dynamic analysis. Michel, Lattion, Oropeza, and Lestuzzi [15] tested two Swiss URM buildings with RC slabs in Visp (Valais) built between 1940 and 1960 using full-scale AVT. The simplified displacement based (DB) approach underestimated the elastic periods leading to too small a displacement capacity. The numerical model gave consistent period values but showed that the severe damage grade may be overestimated in the proposed method. Instead, they proposed a coupling of the methods to obtain all the fragility curves [15].

Michel, Oropeza, and Lestuzzi [16] identified the vulnerability of 21 typical buildings in Visp, Valais, Switzerland, including AVT and analytical model computation. Fragility curves for URM structures with rigid slabs were proposed according to a newly developed method.

Michel et al. [17] found a value of two-thirds of the ambient vibration frequency to be relevant for the earthquake engineering assessment of URMs. As soil–structure interaction affected these parameters, an analytical methodology was proposed to first derive the fixed-base frequency before using these results. Taleb et al. [18] identified and evaluated the seismic properties of an old irregularly shaped three-story URM building in Algiers by ambient and forced vibration tests. The building was rehabilitated by a hysteretic damper system. Anastasia and Athanasios [19] surveyed a post-earthquake scenario in Attica, Greece. In the research, the fundamental periods of vibration for RC buildings were calculated according to existing relationships. They also drew the parameters (height, structural type, and so on) that influence the seismic response and the developed damage based on the wide database. Boutin et al. [20] presented the dynamic characteristics for school buildings in Quebec based on AVTs, which contributed to the development of empirical formulas for the prediction of fundamental vibration periods in URM seismic rehabilitation projects. Beskhyroun, Wotherspoon, and Ma [21] investigated the dynamic behavior (modal parameters utilizing a range of methods in both the frequency and time domain) of a full-scale 13-story RC office building at the University of Auckland under forced vibration, AVT, and distal earthquake excitation. They found consistent results for modal parameters obtained.

This study used the AVT data to identify the predominant frequencies of several URM buildings in Dhaka City with a view to determining the natural structural periods of those buildings. Later, it endeavored to create an equation for deriving structural periods of low-rise URM buildings from the structural parameters of those buildings.

AVT

URM BUILDINGS

The URM buildings under study are located on the northern bank of the Buriganga River in a highly and densely populated metropolitan city that frequently faces earthquakes of low to medium intensity and, based on historical recurrence, that expects some serious seismic threats in the near future. Locations of the buildings are shown in **Fig. 1**.

In Location 1, the soil is light brown to brown or reddish brown or light red to red clay/silt with gray mottling with trace amounts of fine sand having low to high plasticity. In Location 2, the soil is light brown to brown or reddish brown or light red to red clay with gray mottling and trace amounts of considerably fine sand having low to high plasticity. But for Location 3, the soil of the southern part that is nearer to the river is brown or reddish brown or red clay with gray mottling with trace to considerable amounts of fine sand having high plasticity [22]. The soil in Location 4 is generally stiff silty clay.

FIG. 1 Different locations of targeted URMs in Dhaka City.

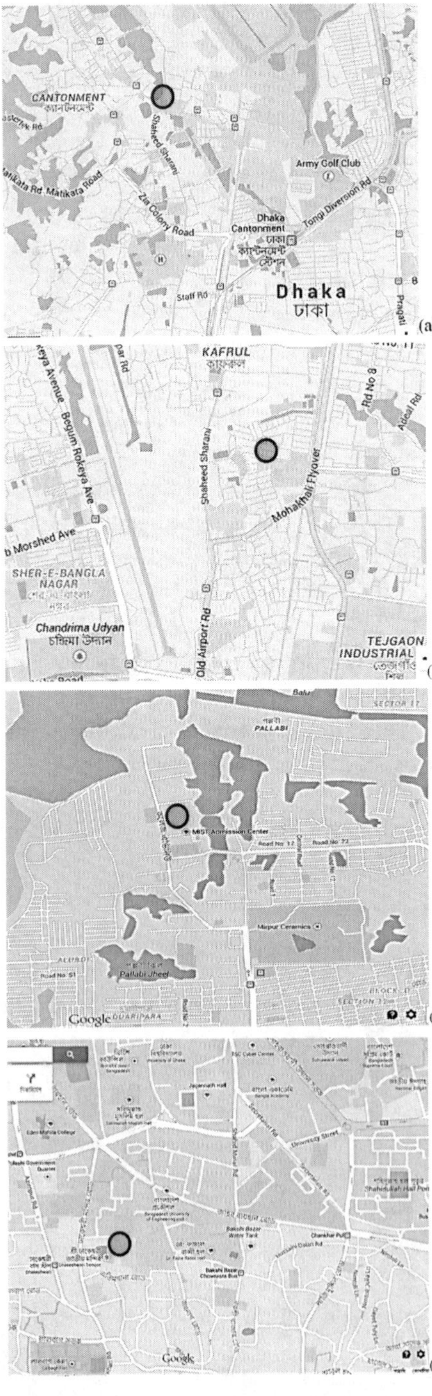

(a) Location 1

(b) Location 2

(c) Location 3

(d) Location 4

All the structures of Location 1 were designed and built in 1960, those of Location 2 in 1972, of Location 3 in 1977, and of Location 4 in 1990. The designs and drawings related to structures built before 1980 could not be found. All data related to those buildings are taken from visual observations. The structures are composed of load-bearing walls of clay bricks in both the longitudinal and transverse direction, with slabs made of RC. The roofs are constructed the same as the floors, providing the possibility for additional stories to be added. The load-bearing walls are solid brick masonry of different thicknesses (25 cm, 39.2 cm, 40 cm, and 43 cm). Bricks are of standard dimensions, 24 cm by 11.43 cm by 6.99 cm, connected by Portland cement mortar. All of the foundations are assumed to be wall foundations with stepped footings. This is based on the design found for the renovation works of all three-story buildings. The width of the bottom step is 1 m and then, in an upward direction, 0.875 m, 0.75 m, 0.625 m, 0.50 m, 0.375 m, and ultimately 0.25 m at the ground level (GL). Side-by-side photos and plan views of the URM buildings are given in **Fig. 2**.

Visual inspection revealed no major cracks in any of the structures. These structures were regulated by construction practices in effect at the time when they were built. It is worth mentioning that there was no authorized national code in effect in Bangladesh at that time.

DATA PROCESSING

The dynamic behaviors of targeted URM buildings were monitored by a Kelunji EchoPro Seismic Recorder. The Kelunji EchoPro Seismic Recorder is a product of Environmental Systems & Services of Australia. Simultaneous recordings in a maximum of four points of any structure are made possible by four triaxial sensors connected with a Micro-ElectroMechanical Systems type accelerometer and cables for the vibration measurements. However, the equipment procured by Bangladesh University of Engineering and Technology (BUET) had three triaxial sensors supplied with it that were available for the research.

The time history records obtained were converted into common text files by Eqwave 3.4 software supplied with the equipment. These were converted into MS Excel files. From the data acquired, data of seven minutes each from the morning, noon/afternoon, and night were gathered together separately for both the top floor and the ground floor. This overlapping was done to increase the number of data points for determining the average. A total of 21 minutes of data were obtained (i.e., $21 \times 6000 = 126{,}000$ separate data points) for the top floor and ground floor that were then converted into a legible plot through a fast Fourier transform (FFT) program created by a MATLAB software package (MathWorks, Inc., Natick, MA). In the process, the program ran based on $212 = 4{,}096$ data points from each one-minute time series. So, there were $21 \times 4096 = 86{,}016$ data points each for the top floor and ground floor accelerations. They were followed by data representing top floor accelerations/ground floor accelerations as a means to transfer function. The transfer function was applied to eliminate the noises at the ground floor and to get different modes of the ambient vibrations in the plot.

FIG. 2 Photos and plan view of Building Nos. (a) 6A, 6B, and 6C; (b) 8; (c) SLA; (d) 13; (e) 10; (f) 15A; (g) 15B; (h) 18A and 18B; (i) 2; (j) 7; (k) 9A and 9D; (l) 11; and (m) 46.

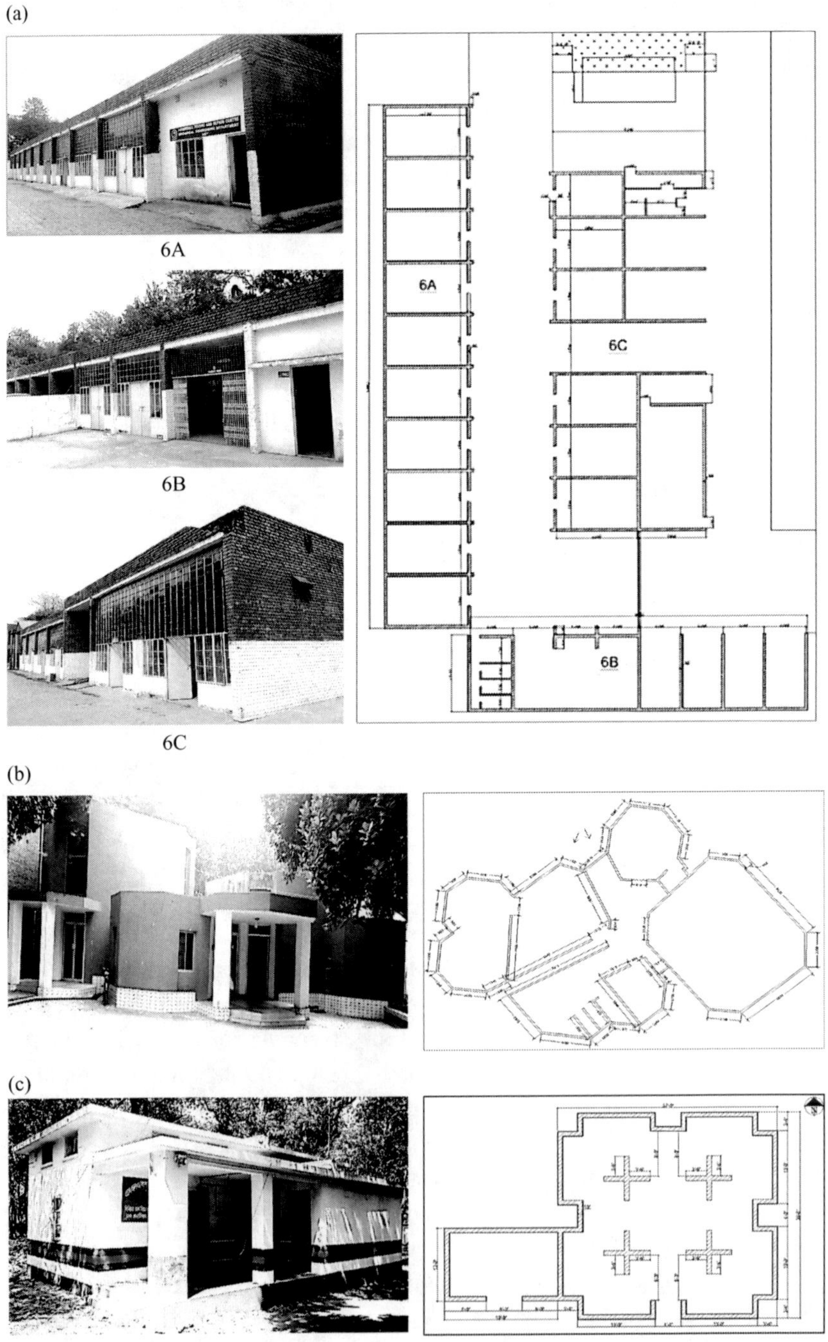

FIG. 2 *(continued)*

(d)

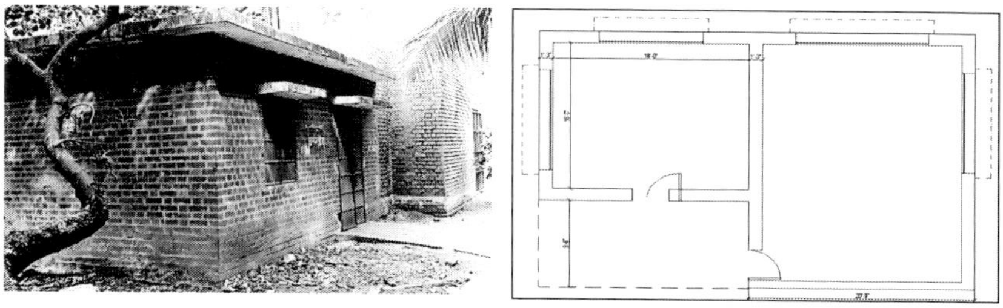

(e)

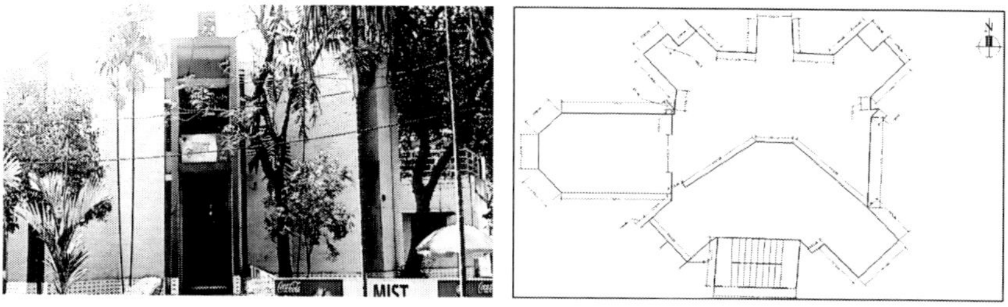

(f)

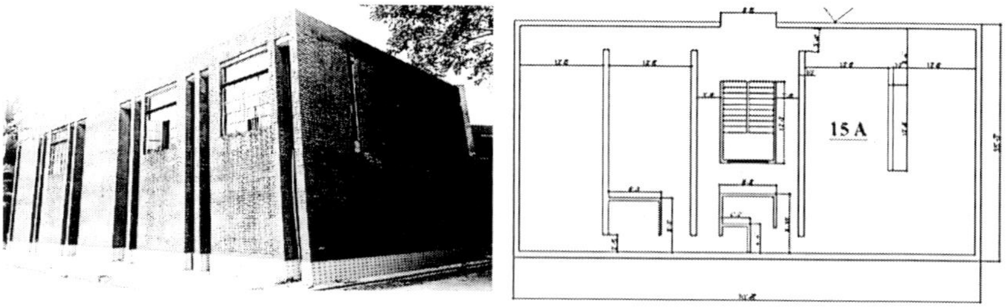

(g)

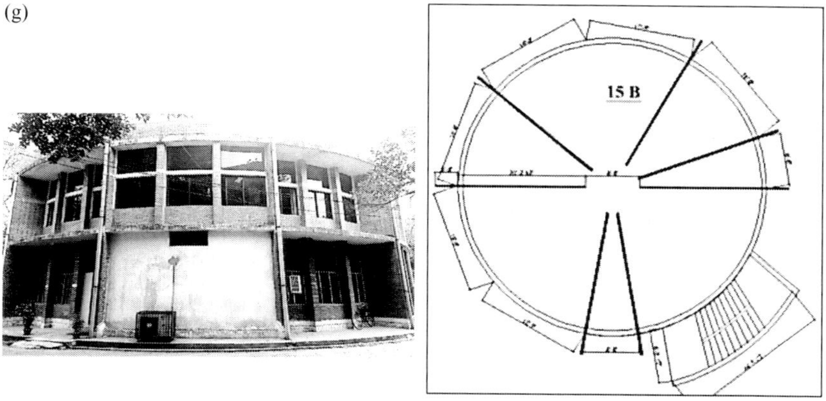

FIG. 2 *(continued)*

(h)

18A 18B

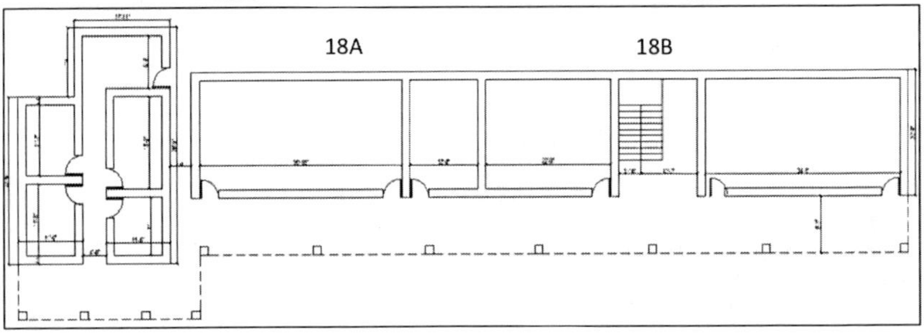

(i)

2

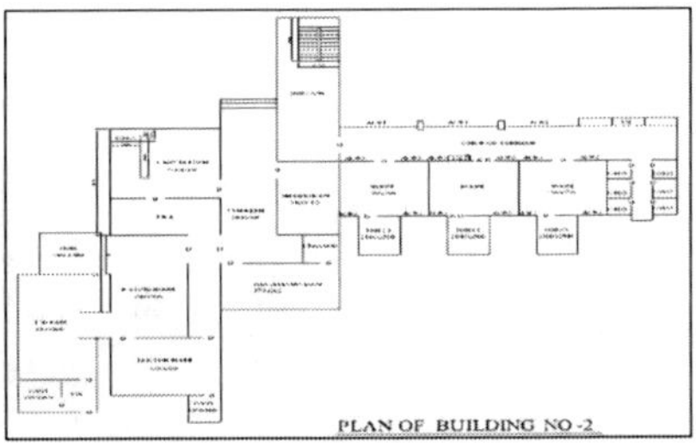

(j)

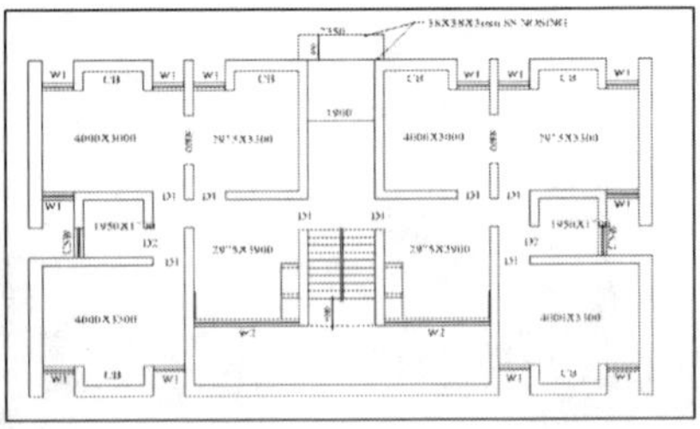

(k)

9A 9D

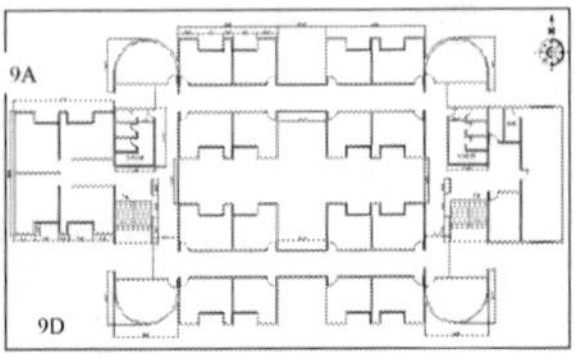

FIG. 2 *(continued)*

(l)

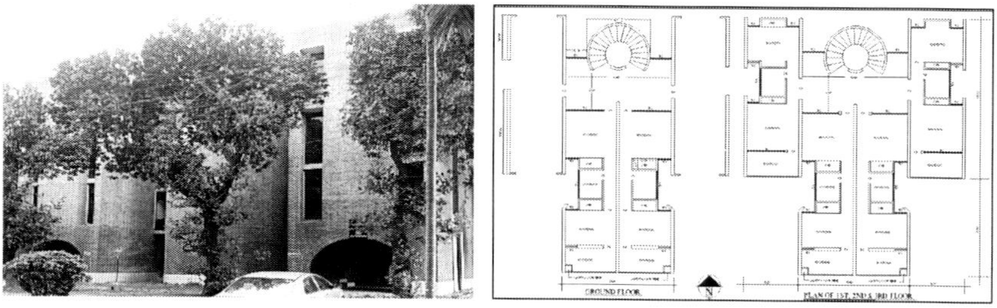

(m)

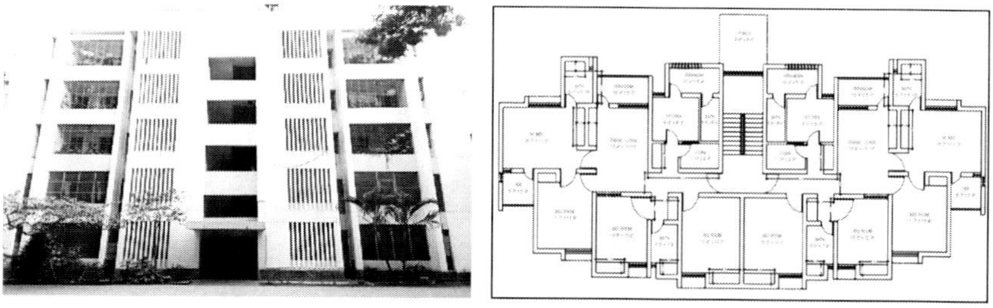

FFT was performed on all the time signals to convert these to the frequency domain. Thus Fourier spectra were composed based on digital FFT. Then the predominant structural vibration frequencies from Fourier spectra were obtained where the magnitudes of the spectral densities were plotted against frequency. From this fundamental frequency, fundamental periods in the longitudinal and transverse directions of the instrument for each URM were acquired. As vertical forces are usually less important in structural engineering, especially SHM, this research focused on the lateral forces for identifying the modal shapes.

AMBIENT MODAL IDENTIFICATION

Ambient vibrations have a random nature and cover a relatively wide band of frequencies [23]. From there, the peaks are identified as resonance frequencies at a particular point. Mode shapes of the structure can then be estimated from those plots using the classical peak picking (PP) method. PP is probably the simplest, yet very useful, quick and practical, system identification method [24]. However, PP has the difficulty of dealing with a close-spaced mode.

AVT RESULTS OF SINGLE-STORY URM BUILDINGS

The Fourier spectra of top floor and ground floor accelerations and modified top floor acceleration (after performing the transfer function) for the north-south direction of Building No. 13 are given in **Fig. 3**.

FIG. 3 FFT output of (a) top, (b) ground, and (c) modified top floor accelerations in north-south direction of Building No. 13.

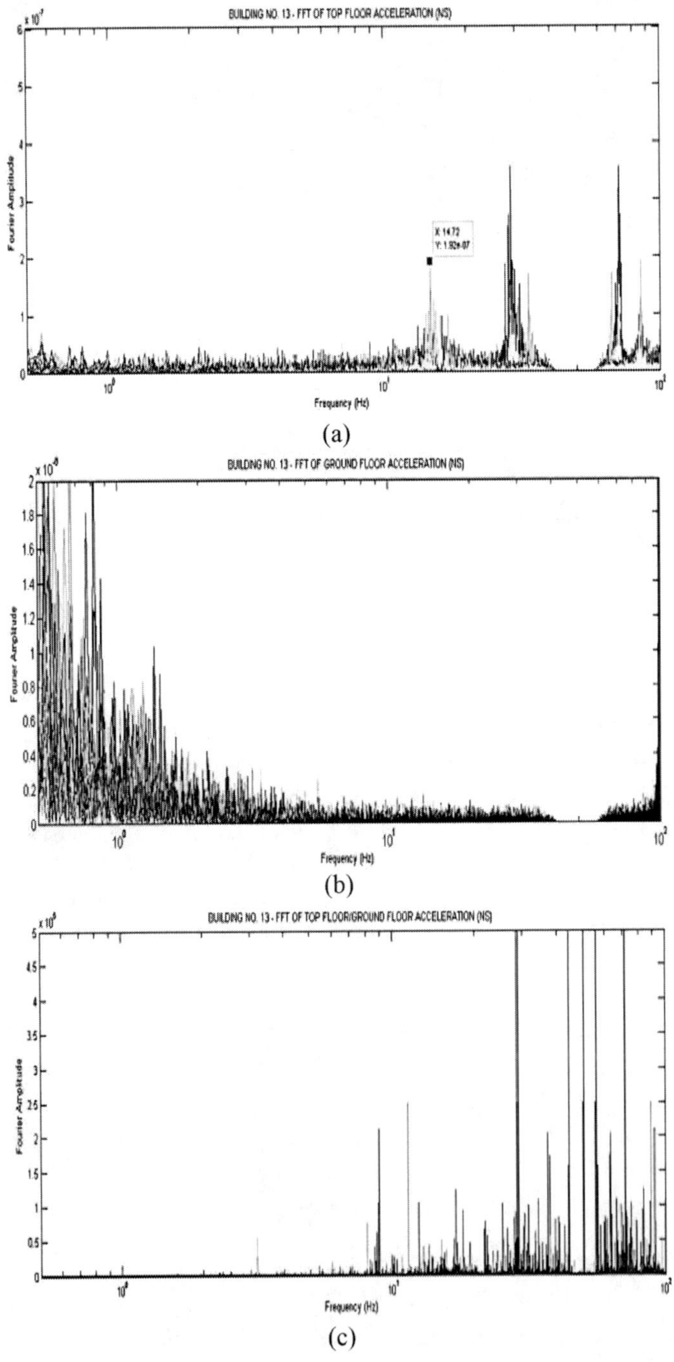

The modes from modified top floor accelerations were not distinct, as shown in Fig. 3, and might lead to erroneous determination of predominant frequencies. Thus, for all the URM buildings under study, predominant frequencies were obtained from the top floor accelerations in both the horizontal cardinal directions. Predominant frequencies of all the targeted URM buildings are given in Table 1.

It is noticeable that the predominant frequencies of the single-storied buildings range from 8.77 Hz to 17.78 Hz irrespective of their heights. Again, the predominant frequencies of the three two-storied buildings range from 6.30 Hz to 10.67 Hz irrespective of their heights and those of the four-storied buildings range from 3.70 Hz to 6.62 Hz irrespective of their heights.

It is worth mentioning that the modal shapes and the predominant frequencies become more and more distinct with an increase in a structure's height.

Results

GENERAL

Fundamental periods of vibration are needed both in the design of new buildings and in the assessment of existing ones so that their seismic response can be evaluated. The knowledge of fundamental frequency and damping ratio of structures is of uppermost importance in earthquake engineering, especially to estimate the seismic demand [25]. Any external force applied to a structure generally is distributed among the three components of the structure, that is, its stiffness, damping, and

TABLE 1 Predominant frequencies of URM buildings.

Stories in the Building	Building No.	Maximum Height at Top Floor from Plinth Level (m)	Dominant Frequency at Top Floor (Hz)	
			NS	EW
1	13	2.59	14.72	15.60
	SLA	3.28	9.75	10.80
	6A		15.10	8.77
	6B	3.81	17.78	11.25
	6C	4.57	9.95	10.28
	8	6.10	11.45	11.52
2	10	6.10	9.77	9.77
	15A		10.67	6.30
	15B	7.32	8.87	7.27
3	2	9.144	4.087	4.087
4	7		5.57	5.32
	11	12.192	6.62	3.70
	9A		5.283	3.833
	9D	14.64	5.133	3.75
5	46	15.24	3.70	3.63

mass components. Hence, structural parameters have great influence on the dynamic properties of structures.

RELATIONSHIP BETWEEN STRUCTURAL PERIODS AND PARAMETERS OF URM BUILDINGS

Structural parameters such as geometry, height of structure, height of story, thickness of load-bearing wall, length of wall in two cardinal directions, RC slab thickness, construction materials, and so on have great effect on the dynamic properties (i.e., the structural period of structures). Initially, it was found that the three-story building gave inconsistent frequency compared to others. Later, it was found that the building had some RC columns in the staircase. Thus, the building was dropped out of the research. First, the study considered the relationship between the length of wall and the frequency of structures irrespective of their other characteristics. As usual, it was found that the frequency increases with the length of walls within a structure only irrespective of their cardinal directions. Thus, a decrease in structural period with an increase in wall length within the same structure is the only result.

ANALYSIS BY MS EXCEL

Relationship between Structural Periods and Heights of URM Buildings

Using Excel-based analysis, a parabolic relationship was found between structural periods and heights of all URM buildings as seen in **Fig. 4**. The best-fit curvature line shows the lower part of a parabola.

R-squared is a statistical measure of how close the data are to the fitted regression line. Zero percent indicates that the model explains none of the variability of

FIG. 4 Relationship between structural periods in north-south directions and heights of structures.

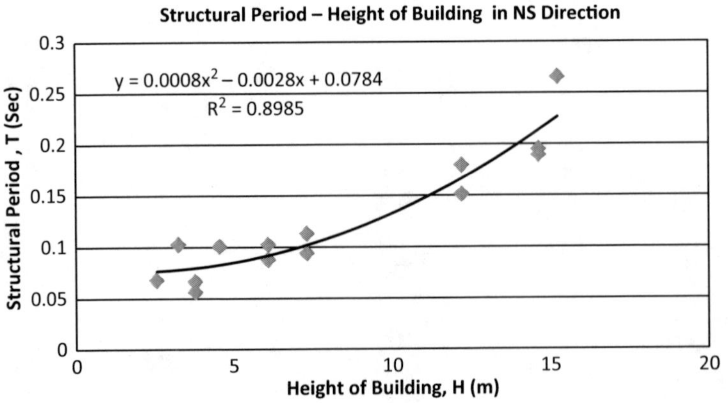

the response data around its mean; 100 % indicates that the model explains all the variability of the response data around its mean. In Fig. 4, the coefficient of determination, R-square value represents the strength of the association between the independent and the dependent variables. The independent variable is strongly associated with the dependent variable.

When considering building orientation in the east-west direction and walls in the north-south direction, the north-south direction is considered as the principal direction. The number of such buildings was ten. A parabolic relationship also exists between structural periods and heights of all ten URM buildings, which is shown in Fig. 5. However, this parabola has a larger radius and a stronger relationship than that of Fig. 4.

In this case, the association of the independent variable with the dependent variable is much stronger.

Again, in the east-west direction, there is a parabolic relationship found between structural periods and heights of all URM buildings irrespective of the principal directions of the structures, as seen in Fig. 6. However, the best-fit curvature line also shows the lower part of a parabola.

The ten URM buildings considered in Fig. 5 have their principal axis oriented in the north-south direction. When the relationship between their structural periods and heights is considered in the east-west direction, a different parabolic relationship with a strong association between the independent variable and the dependent variable is obtained. This relationship is shown in Fig. 7, which has a stronger relationship than that of Fig. 6.

It also was found that five URM buildings having a rectangular-shaped plan have the similar strong parabolic relationship between their structural periods and heights in both directions.

FIG. 5 Relationship between structural periods in the principal directions of structures (north-south oriented) and heights.

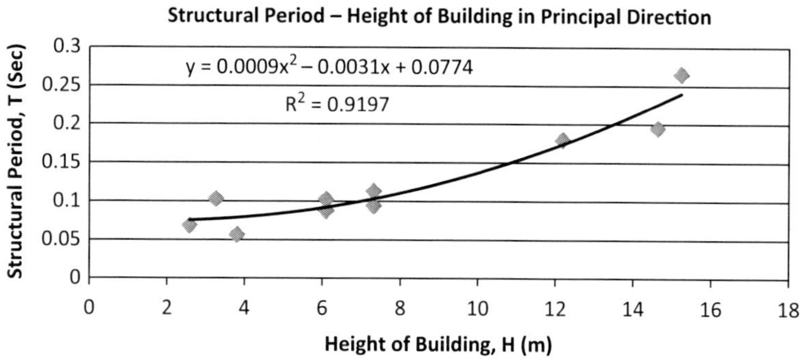

FIG. 6 Relationship between structural periods in east-west direction and heights of structures.

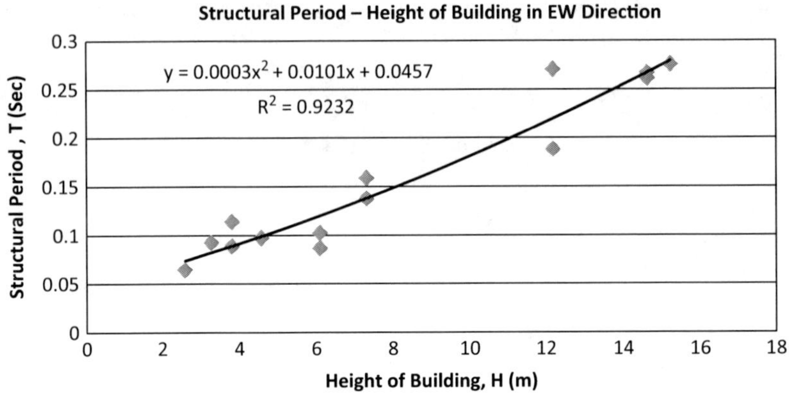

FIG. 7 Relationship between structural periods and heights in the east-west (opposite to principal) directions of structures.

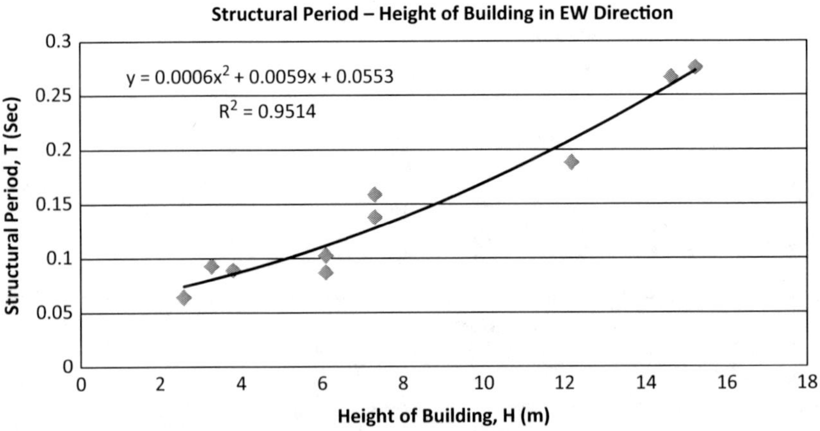

RELATIONSHIP BETWEEN STRUCTURAL PERIODS AND TOTAL LENGTHS OF WALL OF URM BUILDINGS

The Excel-based analysis shows that there is a parabolic relationship found between structural periods and walls of URM buildings irrespective of the principal directions of the structures. However, the best-fit curvature line shows the upper part of a parabola. Again, the linear plot represents almost a similar strength of relationship ($R^2 = 0.76$ approximately).

Considering walls in the north-south direction of a building, a similar but stronger parabolic relationship ($R^2 = 0.90$ approximately) exists between structural periods and walls of ten URM buildings, representing the best-fit curvature line in an upper part of a parabola. The linear plot represents almost a similar strength of relationship but is impractical.

Again, in the east-west direction, a parabolic (upper part) relationship was found between structural periods and lengths of URM buildings irrespective of principal directions of the structures. The linear plot represents a somewhat lesser strength of relationship ($R^2 = 0.71$ to 0.73 approximately).

In the direction opposite to the principal direction (north-south direction with their longer axis), in an overall perspective, the association of the independent variable with the dependent variable is stronger. The linear plot represents a somewhat lesser strength of relationship ($R^2 = 0.73$ to 0.74 approximately).

It is observed that structures with great total wall lengths may have lesser structural periods than those with less total wall lengths. However, this relationship may not be the case in reality. It is obvious that the structural period increases with the increase in total wall length. So this parabolic relation of structural period with total wall length of structures can be ignored. Since the linear relationship has the practical significance, it may be accepted.

RELATIONSHIP BETWEEN STRUCTURAL PERIODS AND ASPECT RATIO OF URM BUILDINGS

The relationships show that a parabolic relation exists between structural periods of both the cardinal direction and the aspect ratios of the structures. However, when considering the R-square value, the strength of association between the dependent and independent variables is very weak ($R^2 = 0.22$ approximately).

RELATIONSHIP BETWEEN STRUCTURAL PERIODS AND RATIO OF THE HEIGHTS AND THE LEAST DIMENSIONS IN THE PLANS OF URM BUILDINGS

The relationships show that a parabolic relation exists between structural periods of both the cardinal direction with the ratio of the heights and the least dimensions in the plans of the structures. The R-square values of both these relationships do not represent convincing strength of association between the dependent and independent variables ($R^2 = 0.67$ to 0.79 approximately).

ANALYSIS OF THE RESULTS

Results Obtained from the Formula of Bangladesh National Building Codes (BNBC)

The structural periods T (Sec) of ten URM buildings were obtained from BNBC 1993 from the following (Eq 1):

$$T = C_t(h_n)^{3/4} \tag{1}$$

where:

$C_t = 0.049$ for all other structural systems, and

h_n = height in meters above the base to level n.

Again, the same was obtained from Draft BNBC 2015 from Eq 2:

$$T = C_t(h_n)^{\mathrm{m}} \tag{2}$$

where h_n is the height of building in meters from the foundation or from the top of a rigid basement. This excludes the basement stories, where basement walls are connected with the ground floor deck or fitted between the building columns. But, it includes the basement stories, when they are not so connected; C_t and m are 0.488 and 0.75, respectively.

Results Obtained from MS Excel Plots

Ten URM buildings are oriented in the north-south direction, and their structural periods T (Sec) with their heights are related as follows (Eq 3):

$$\mathrm{T} = 0.0009\mathrm{H}^2 - 0.0031\mathrm{H} + 0.0774 \tag{3}$$

Again, the same URM buildings are related with their structural periods T (Sec) and total length of walls as follows (Eq 4):

$$\mathrm{T} = 0.0005\mathrm{L} + 0.0574 \tag{4}$$

Results Obtained from SPSS 11.5

The Statistical Package for the Social Sciences (SPSS) is a software package used for logical batched and non-batched statistical analysis. Originally produced by SPSS Inc. (Chicago, IL), it was acquired by IBM in 2009. Correlation among different factors can be made through regression analyses.

Taking the experimental data of the north-south direction, the study performed analyses of twelve models (four models each—enter, stepwise, backward elimination, and forward selection procedure of regression) of all three approaches. It was found that only a single model is suitable to be used as the relation among structural periods, height, and the total length of walls of the structures (i.e., $\mathrm{T} = 0.071 + 0.001\ \mathrm{H}^2 + 3.039\mathrm{E}\text{-}07 * \mathrm{L}^2$). Again, Approach 1 and Approach 2 both support it being statistically and practically significant ($\mathrm{R}^2 = 0.957$, Adjusted $\mathrm{R}^2 = 0.945$, and $\mathrm{SE} = 0.0155$). The same procedure was followed by taking the experimental data of the east-west direction. However, no model could be accepted as all twelve models failed to provide any relation combining the height and the total length of walls with structural periods of the structures.

The coefficient of the third term of the equation is 3.039E-07 (i.e., 0.0000003039), which is very insignificant when it is multiplied with L^2. It may carry some contribution for the structural periods of buildings having great total length of walls in meters.

COMPARISON AMONG THE RESULTS OBTAINED

A comparative statement of all values of structural periods obtained from different formulas and procedures is provided in Table 2.

Except for Building no. SLA, the general observations are stated as follows:

- The structural periods obtained using the formulas of both the BNBCs are much higher than the experimental values and values calculated from the MS Excel parabolic equation of the same. It signifies that BNBC states larger structural periods, which leads to determination of lesser design base shear and thus designing unsafe URM structures.
- Again, the structural periods obtained using the SPSS 11.5 are a little higher than the experimental values and the values calculated from the MS Excel parabolic equation of the same.
- The structural periods obtained using the MS Excel plots are very close to those of the experimental values. Average variation between the values of the MS Excel plot and experimental values is 13.3 %.

FINAL DEDUCTION

The coefficient of determination, R^2, is the proportion of variance in the dependent variable that can be explained by the independent variables. This is an overall measure of the strength of association between the independent variable and the dependent variable. The larger the value, the better the regression line describes the data. It is worth mentioning and noticing that the MS Excel polynomial fit (i.e., parabolic fit) for the relation between structural periods, T (Sec), with the heights, H (m), of structures has the R^2 value of 0.9197. However, the MS Excel linear fit for the relation between structural periods, T (Sec), with the total length of walls, L (m), of structures, has the R^2 value of 0.8923.

Therefore, keeping in mind all the facts and findings of the research, Eq 3 may be utilized to obtain the structural periods, T (Sec), while taking the height, H (m),

TABLE 2 Comparative statement of the structural periods.

Building No.	H (m)	BNBC 1993	Draft BNBC 2015	Experiment	MS Excel Plot for Best Fit Curve		SPSS Regression
					T vs H	T vs L	
13	2.59	0.1000394	0.09963	0.067935	0.075408	0.074825	0.07772
SLA	3.28	0.1194267	0.11894	0.102564	0.076915	0.068795	0.08179
6B	3.81	0.1336256	0.13308	0.056243	0.078653	0.083145	0.08558
8	6.10	0.1901925	0.18942	0.087336	0.091979	0.09667	0.10863
10	6.10	0.1901925	0.18942	0.102354	0.091979	0.09113	0.10863
15A	7.32	0.2180617	0.21717	0.093721	0.102932	0.11460	0.12545
15B	7.32	0.2180617	0.21717	0.112740	0.102932	0.12979	0.12545
7	12.192	0.3197071	0.31840	0.179533	0.173385	0.17878	0.22636
9B	14.64	0.3667346	0.36524	0.194932	0.224913	0.167375	0.29929
46	15.24	0.3779504	0.37641	0.265252	0.239188	0.299925	0.31965

of structures into consideration. So the study suggests a change in the present code (BNBC 1993) for calculating the structural periods of URM structures ranging from single-storied to five-storied (height ranges from 2.59 m to 15.24 m) by using Eq 3 in the following form (Eq 5):

$$T = 0.001H^2 - 0.003H + 0.08 \tag{5}$$

Conclusion

The actual responsive behavior of structures undergoing an earthquake is similar to the ambient vibration behavior. That is why it is being used for SHM. The research included both software analysis and physical testing of URM buildings. Software analysis using accurate nonlinear time history analysis of AVT enabled researchers to analyze the URM buildings with their ambient vibrations and geometry.

The structural periods obtained using both formulas of the BNBC are much higher than the experimental values and values calculated from the MS Excel parabolic equation of the same. Again, the structural periods obtained using the SPSS 11.5 are a little higher than the experimental values. However, the structural periods obtained using the MS Excel plots are very close to those of the experimental values.

FINDINGS FROM THE STUDY

An attempt was made to correlate structural periods with heights and structural periods with total wall length of structures by MS Excel plots. Again, among all the three parameters (i.e., structural periods, heights, and total wall lengths of structures), a correlation was modeled by various methods of SPSS. Finally, a model was selected combining structural periods, heights, and total wall lengths of structures. However, that statistically significant model has very little practical significance.

Bearing in mind all the facts, the following findings from the research can be summarized:

a. The modal shapes and the predominant frequencies become more and more distinct with the increase of a structure's height.
b. There is a strong relationship among structural periods, heights, and total length of walls in two cardinal directions of URM structures.
c. There is a strong parabolic relationship among the structural periods and the heights and linear relationship between the structural periods and the total length of walls in two cardinal directions of URM structures. However, the relationship between the structural periods and the heights of URM structures is much stronger.
d. The structural periods obtained using both formulas of the BNBC are much higher than the experimental values and values calculated from the MS Excel parabolic equation and SPSS models of the same. This signifies that BNBC states larger structural periods, which leads to determination of lesser design base shear and thus designing unsafe URM structures.

e. The structural periods obtained using the SPSS model are a little higher than the experimental values and the values calculated from the MS Excel parabolic equation of the same.
f. The structural periods obtained using the MS Excel parabolic equation are very close to those of the experimental values.

Based on all of the findings, the study proposes that Eq 5 may be utilized to obtain the structural periods, T (Sec), while taking height, H (m), of structures into consideration. The study suggests a change in the present BNBC 1993 for calculating the structural periods of URM structures ranging from single-story to five-stories (height ranges from 2.59 m to 15.24 m).

RECOMMENDATION FOR FUTURE RESEARCH

The following are the recommendations for future study.

- If the number of sample URM buildings and locations are increased within Dhaka City or in a bigger frame countrywide, it is likely to provide a validation of this study. A period database can then be developed for URM buildings in Dhaka City or, for that matter, for URM buildings in Bangladesh.
- Additional research could be conducted to more deeply examine this field by overcoming the limitations presented earlier.
- More design variables (structural parameters such thickness of walls, thickness of slabs, types of foundations, aspect ratio, slenderness ratio, age of structures, variations in construction materials, and so on) may be included in the future.
- Other nonlinear models and software can be used to conduct further research.
- Similar research can be conducted on structures other than URM buildings (i.e., arch bridges, chimneys, towers, and so on).
- It is also expected that the modeling technique will unveil a new avenue for the researchers of Bangladesh when making further studies of a similar nature.
- Further research based on AVT can be conducted using models prepared in the laboratory. Work using a laboratory shake table testing model can be important.
- Very important masonry structures (i.e., URM KPIs of Bangladesh) may be instrumented permanently for their SHM through continuous recordings of ambient vibrations.

ACKNOWLEDGMENTS

All praises to the Creator of the world for His kindness and blessings for giving the author the opportunity, determination, and strength to do this work. The author wishes to express sincere appreciation and deepest gratitude to Dr. Raquib Ahsan for his continuous and patient support, guidance, and encouragement. His supervision inspired the author to complete the study by overcoming all persistent pediments of his professional environment. Without Dr. Ahsan's honest motivation as well as caring attitude, it would not have been possible for the author to have completed this massive task.

References

[1] Ministry of Public Works, House Building Research Institute and Bangladesh Standards and Testing Institute, "Bangladesh National Building Code," Bangladesh Government (B.G.) Press, Tejgaon, Dhaka, Bangladesh, 1993.

[2] Ivanovic, S. S., Trifunac, M. D., and Todorovska, M. I., "Ambient Vibration Tests of Structures—A Review," *ISET J. Earth. Technol.*, Vol. 37, No. 4, 2000, pp. 165–197.

[3] Farrar, C. R. and James, G. H., III, "System Identification from Ambient Vibration Measurements on a Bridge," *J. Sound Vib.*, Vol. 205, No. 1, 1997, pp. 1—18.

[4] Clinton, J. F., Bradford, S. C., Heaton, T. H., and Favela, J., "The Observed Wander of the Natural Frequencies in a Structure," *B. Seismol. Soc. Am.*, Vol. 96, No. 1, 2006, pp. 237–257.

[5] Carder, D. S., "Observed Vibrations of Buildings," *B. Seismol. Soc. Am.*, Vol. 26, No. 3, 1936, pp. 245–277.

[6] Crawford, R. and Ward, H. S., "Determination of the Natural Periods of Buildings," *B. Seismol. Soc. Am.*, Vol. 54, No. 6, 1964, pp. 1743–1756.

[7] Ward, H. S. and Crawford, R., "Wind-Induced Vibrations and Building Modes," *B. Seismol. Soc. Am.*, Vol. 56, No. 4, 1966, pp. 793–813.

[8] Klopp, G. and Griffiths, M. C., "Dynamic Characteristics of Existing Masonry Buildings," *Proceedings of the Pacific Conference on Earthquake Engineering*, Auckland, New Zealand, November 20–23, 1991, Vol. 2, 1991, pp. 187–195.

[9] Balkaya, C. and Kalkan, E., "Estimation of Fundamental Periods of Shear-Wall Dominant Building Structures," *Earthq. Eng. Struct. D.*, Vol. 32, No. 7, 2003, pp. 985–998.

[10] Turek, M., Thibert, K., Ventura, C., and Kuan, S., "Ambient Vibration Testing of Three Unreinforced Brick Masonry Buildings in Vancouver, Canada," *Project Report for the Vancouver Chinatown Revitalization Committee*, 2004.

[11] Gentile, C. and Saisi, A., "Ambient Vibration Testing of Historic Masonry Towers for Structural Identification and Damage Assessment," *Constr. Build. Mater.*, Vol. 21, No. 6, 2007, pp. 1311–1321.

[12] Gilles, D. and McClure, G., "Development of a Period Database for Buildings in Montreal Using Ambient Vibrations," presented at the 14th World Conference on Earthquake Engineering (WCEE), Beijing, China, October 12–17, 2008.

[13] Hancilar, U., Durukal, E., and Erdik, M., "Seismic Assessment and Rehabilitation of A Historical Unreinforced Masonry (URM) Building in Istanbul," presented at the 14th World Conference on Earthquake Engineering (WCEE), Beijing, China, October 12–17, 2008.

[14] Crowley, H. and Pinho, R., "Revisiting Eurocode 8 Formulae for Periods of Vibration and Their Employment in Linear Seismic Analysis," *Eurocode 8 Perspectives from the Italian Standpoint Workshop*, E. Cosenza, Ed., Doppiavoce, Napoli, Italy, 2009, pp. 95–108.

[15] Michel, C., Lattion, E., Oropeza, M., and Lestuzzi, P., "Vulnerability Assessment of Existing Masonry Buildings in Moderate Seismicity Areas Using Experimental Techniques,"

Proceedings of the 2009 Asian-Pacific Network of Centers for Earthquake Engineering Research (ANCER) Workshop, Urbana-Champaign, IL, August 13–14, 2009, Paper #142.

[16] Michel, C., Oropeza, M., and Lestuzzi, P., "Validate the Building Inventory and Important Infrastructure in the Test Area," Applied Computing and Mechanics Laboratory, EPFL, Coupled Seismogenic Geo-hazards in Alpine Regions (COGEAR), Competence Center Environment and Sustainability (CCES), January 27, 2010.

[17] Michel, C., Zapico, B., Lestuzzi, P., Molina, F. J., and Weber, F., "Quantification of Fundamental Frequency Drop for Unreinforced Masonry Buildings from Dynamic Tests," *Earthq. Eng. Struct. D.*, Vol. 40, No. 11, 2011, pp. 1283–1296.

[18] Taleb, R., Bouriche, F., Remas, A., Boukri, M., and Kehila, F., "Use of Ambient and Forced Vibration Tests to Evaluate Seismic Properties of an Unreinforced Masonry Building Rehabilitated by Dampers," *Proceedings of the 15th World Conference on Earthquake Engineering (WCEE)*, Lisbon, Portugal, September 24–28, 2012, Curran Associates, Inc., Red Hook, NY, 2015, pp. 1231–1240.

[19] Anastasia, K. E. and Athanasios, I. K., "Correlation of Structural Seismic Damage with Fundamental Period of RC Buildings," *Open J. Civil Eng.*, Vol. 3, No. 1, 2013, pp. 45–67.

[20] Boutin, M., Proulx, J., Mestar, M., Nollet, M., Tischer, H., and McClure, G., "Dynamic Characterization of URM School Buildings in Quebec," *Proceedings of the 12th Canadian Masonry Symposium*, Vancouver, British Columbia, June 2–5, 2013, Paper No. 485.

[21] Beskhyroun, S., Wotherspoon, L., and Ma, Q. T., "System Identification of A 13-Story Reinforced Concrete Building through Ambient and Forced Vibration," *Proceedings of the 4th International Conference on Computational Methods in Structural Dynamics and Earthquake Engineering (COMPDYN)*, Kos Island, Greece, June 12–14, 2013, M. Papadrakakis, V. Papadopoulos, and V. Plevris, Eds., National Technical University of Athens, Athens, 2014, pp. 3448–3458.

[22] Serajuddin, M., H., M. A., Hossain, M. J. and Islam, M. M. (2001), "Characterization of Uplifted Pleistocene Deposits in Dhaka City," *J. Civ. Eng.*, Vol. 29, No. 2, 2001, pp. 167–186.

[23] Vlad, I. and Vlad, M., "Ambient Vibration Instrumental Investigations on Full Scale Structures," *Proceedings of the 8th International Conference on Structural Dynamics, EURODYN 2011*, Leuven, Belgium, July 4–6, 2011, Katholieke Universiteit Leuven, Leuven, Belgium, 2011, pp. 2424–2431.

[24] Omenzetter, P. R, Beskhyroun, S., Shabbir, F., Chen, G.-W., Chen, X., Wang, S., and Zha, A., "Forced and Ambient Vibration Testing of Full Scale Bridges," Earthquake Commission Research Foundation, Wellington, New Zealand, 2013, Project No. UNI/578.

[25] Reaz, M., "Feasibility of Using Dampers for Reducing Earthquake Damage in Buildings," Investigative Research Project, Newcastle University, Newcastle upon Tyne, UK, 2012.

STP 1612, 2018 / available online at www.astm.org / doi: 10.1520/STP161220170153

Zack Olson,[1] Wes Cook,[2] and Dale Henneke[2]

Sustainable Cellulose Masonry

Citation

Olson, Z., Cook, W., and Henneke, D., "Sustainable Cellulose Masonry," *Masonry 2018, ASTM STP1612*, N. V. Krogstad and W. M. McGinley, Eds., ASTM International, West Conshohocken, PA, 2018, pp. 56–66, http://dx.doi.org/10.1520/STP161220170153[3]

ABSTRACT

A novel masonry material containing postindustrial cellulose, bentonite clay, and small amounts of other additives (e.g., antifungal, desiccant and water proofing agents) was made and characterized. The overall mechanical properties of the composite material are controlled and tailored by varying the ratio of the primary constituents (postindustrial cellulose and bentonite clay). This grade of cellulose is no longer suitable for use in the manufacture of recycled paper products and, as such, it is considered a waste product. The described process repurposes this cellulose waste as a component in a structural element that can be worked using standard woodworking tools. The material's compressive strength ranges from approximately 338 psi (2.33 MPa) at low bentonite ratios (less than 50 % by mass) to approximately 1,015 psi (7.01 MPa) at higher bentonite ratios (greater than 60 % by mass). The higher bentonite ratio, Ratio B, has a compressive strength similar to that of a Class 2 burnt brick but does not require the material to be fired (i.e., a significant cost and energy savings). This material can be easily painted, stained, or surface-treated much like wood. However, unlike wood and other masonry products (e.g., concrete), it exhibits a failure mode more similar to that of a polymer where the material does not catastrophically fail—rather it slowly yields and shows visible deformation. This mode of failure provides ample warning before a piece fails and gives time for it to be repaired or replaced before catastrophic failure. The material is also lighter

Manuscript received September 8, 2017; accepted for publication February 13, 2018.

[1]New Mexico Institute of Mining and Technology, 21 Rusti Fields Way, Port Angeles, WA 98362, USA http://orcid.org/0000-0002-0258-4412

[2]New Mexico Institute of Mining and Technology, Civil Engineering Dept., 801 Leroy Pl., Socorro, NM 87801, USA W. C. http://orcid.org/0000-0002-7269-0018, D. H. http://orcid.org/0000-0003-0830-2719

[3]ASTM International Symposium on *Masonry 2018* on June 26, 2018 in San Diego, CA, USA.

and cheaper to produce than most competing products as one of its main components is a postindustrial waste product.

Keywords
sustainable, brick, masonry, cellulose, postindustrial material

Introduction

In the United States, the average person generates approximately 4.3 lbs of waste per day [1]. As a country, we generate approximately 220 million tons of waste per year, over half of which ends up in landfills. Landfill waste is the second largest human-caused producer of methane [1], which is a known greenhouse gas. Of this landfill waste, paper waste is one of the largest contributors to municipal waste and, while not the largest, a significant contributor to industrial waste [2,3]. The transport and disposal of cellulose waste on an industrial level is a large problem for manufacturing companies. A significant waste stream of the paper industry is unfit paper pulp. This waste paper pulp is cellulose material generated from the processing of paper products such as cardboard and writing paper as well as a byproduct of the paper recycling process. **Fig. 1** shows the paper materials that do not meet the quality standards for the paper-based products and that are unsuitable for recycling. Discarding and disposal of waste pulp is an expense for the recycling and paper industries. In an effort to alleviate disposal costs and reduce landfill demand, this study explores the use of waste pulp as a primary constituent of this ecologically friendly brick.

FIG. 1 As-received paper pulp.

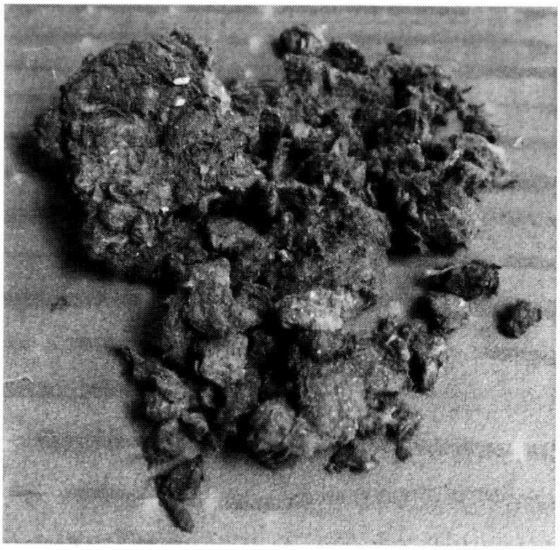

The two main types of conventional bricks used in construction are sun-dried bricks and fired bricks. Sun-dried bricks are dried using solar energy and are typically only used for temporary structures in areas with light rainfall. Fired bricks can be further classified into four categories: First class brick, second class brick, third class brick, and fourth class brick. First class bricks are table molded, have smooth sharp edges, and are fired in kilns. Typically, they are used for permanent structures with an average compressive strength of 1500 psi (10.3 MPa). Second class bricks are made in earthen or ground molds and are also typically fired in kilns. Second class bricks have rougher faces than first class bricks, can feature slight irregularities in shape, and may exhibit cracking on their surfaces. These bricks generally are used in cases where the brickwork is to be covered with a layer of plaster or other surfacing agent, and they have a compressive strength of approximately 1,000 psi (6.9MPa). Third class bricks are again traditionally ground-molded but, unlike second class bricks, are burnt using the brick clamp method. These bricks are irregular in shape and are used primarily in the fabrication of temporary structures in areas that receive little rainfall. Third class bricks have an average compressive strength of 500 psi (3.4 MPa). Finally, fourth class bricks are bricks that have been over burnt, and they are exceedingly irregular in shape and color. Due to their irregularity, they are most often ground into aggregate for use in concrete foundations [4]. The material investigated in this study seeks to directly confront environmental issues by recycling previously unused waste cellulose into a useful structural material. By so doing, landfill demand can be reduced, greenhouse gas emissions can be decreased, and a structural material suitable for masonry construction is the final product.

Evidence that the brick manufactured and evaluated in this study will be ecologically friendly is demonstrated through the constituents and processing discussion. Sample preparation and variation is described along with the experimental protocol. Mechanical properties of uniaxial compressive strength and modulus of elasticity of the brick are assessed and compared to each other and to conventional brick types. A coating system is also tested to evaluate water resistance. Future directions with applications are presented. Finally, conclusions are drawn regarding the suitability of the brick as a structural material based on the mechanical properties determined.

Brick Constituents and Processing

Brick constituents include processed post-consumer cardboard, bentonite clay, calcium chloride, and potable water. **Fig. 1** shows the post-consumer cardboard obtained from the waste stream of a regional cardboard recycling facility. The post-industrial cellulose used in this study was supplied by FS Enterprises [5]. Post-consumer cardboard is pulverized to obtain the material shown in **Fig. 2**. **Fig. 2** shows the fibers post-processing as compared directly to the preprocessed material shown in **Fig. 1**. The processed cellulose is much less compressed and significantly more aerated, increasing the volume of the raw material. **Fig. 3** shows a microscopic

FIG. 2 Processed paper pulp.

FIG. 3 Microscopic image of processed cellulose fiber.

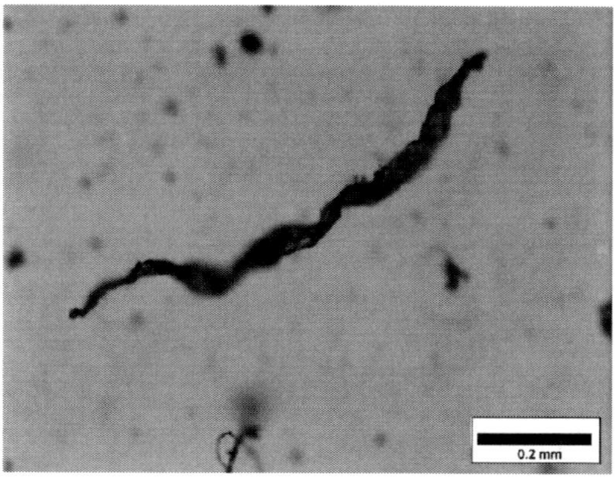

image of the processed cellulose fibers. The helical shape of the cellulose fibers assist in explaining the macroscopic behavior in terms of phenomena at the microscopic level. This image is shown to demonstrate the helical shape of the fibers. Improvements upon an existing patent [5] included pulverizing the cardboard press cake and altering the mixture ratios. Bentonite clay and calcium chloride were obtained from a local retailer. Mixing ratios are based on weight percentages. Fibers are mixed in different weight percentages with bentonite clay while maintaining calcium chloride at 10 %.The fiber-clay ratios spanned from fiber-rich mixtures to clay-rich mixtures. Density across the ratio spectrum increased with increasing clay content as shown in **Table 1**.

Preparation of Test Samples

The test samples were batched according to the ratios shown in **Table 2** with potable water using a food processor. After blending the materials into a paste-like substance, they were then pressed into cylindrical PVC molds 1.5 in. in diameter and 3 in. tall and dried for 28 days. Compaction of cylinders was manually achieved through visual inspection of the fully saturated matrix such that water would bleed

TABLE 1 Elastic modulus and compressive strength.

	Ratio A	Ratio B	Ratio E	Ratio F	Ratio U
Elastic Modulus (psi)					
1	26135.8	30095.4	25207.6	3710.0	14895.4
2	15228.9	57377.0	34954.2	3524.4	12748.8
3	-	-	36839.7	2045.0	12647.3
4	10138.1	60480.8	39740.4	4191.6	6497.7
Average	17167.6	49327.4	33228.2	3379.4	11704.6
0.2 % Offset Yield Strength (psi)					
1	219	900.7	266.9	100.1	147.9
2	-	858.6	104.4	59.5	146.5
3	-	-	297.3	71.1	156.6
4	306	749.8	248.0	108.8	213.2
Average	262.5	836.9	229.2	84.1	165.3
Ultimate Strength (psi)					
1	-	1048.6	-	-	-
2	-	1032.7	-	-	-
3	-	-	-	-	-
4	-	964.5	-	-	-
Average	NA	1016.7	NA	NA	NA
Average Density (lb/in^3)		0.0580	0.0322	0.0193	0.0237

TABLE 2 Comparison of different ratios.

	Component Ratio (Weight %)		
	Bentonite Clay	Press Cake Fiber	Calcium Chloride
Ratio B	75	15	10
Ratio E	50	40	10
Ratio U	30	60	10
Ratio A	15	75	10
Ratio D	5	85	10

out. All samples were air dried at room temperature (70°F) for the 28-day duration. After drying, samples were removed from molds and visually evaluated for any obvious defects. Any samples that failed this inspection were rejected and replaced. Shrinkage of the specimens was observed vertically and radially with variation between ratios. Compressive strength sample preparation included sanding of end faces to a smooth finish for uniaxial testing. The dimensions of the cylinders were measured using dial calipers. Specimen diameters were determined from averaging midpoint and end diameters with additional measurements taken from the upper and lower quarters. The average diameter was then taken as the average of those seven measurements. The length for each sample was also taken using calipers—again with seven measurements being used to form an average. Samples used for the submersion test were coated with the manufacturer's recommended number of layers (two) and followed the recommended dry time between layers (24 h) of siloxane. Once the coating substance was applied, the mass of each sample was measured.

Test Protocol, Observations, and Data

The experimental protocol for uniaxial compression tests follows Section 7 of ASTM C67-09, *Standard Test Methods for Sampling and Testing Brick and Structural Clay Tile* [6]. Samples were prepared as described earlier. Each sample was individually tested and recorded at a load rate in accordance with ASTM C67-09, Section 7. Average compressive strengths are shown for the varying mixture ratios in **Table 1**. Ratio B (75 wt.% clay, 15wt.% fiber, 10 wt.% calcium chloride) had the highest compressive strength, which was 7 MPa (1015 psi), and an elastic modulus of 340 MPa (49.3 KSI). From **Table 1**, it can be observed that by combining the compressive strength of bentonite clay with the tensile strength of cellulose fibers from postindustrial cardboard, a sustainable building material is fabricated that is stronger than clay alone. Additionally, the ecologically friendly brick matches or exceeds the compressive strength requirements for a second class fired brick of 1,000 psi (6.9 MPa).

When tested to failure, it was noticed that the higher percent clay specimens, such as Ratio B, had a higher ultimate compressive strength. The overwhelming majority of samples failed in shear. None of the tested ratios failed in a sudden catastrophic manner that is typical of ceramic materials such as concrete and standard brick. This observation means that the material exhibits a slower more visible failure mode than standard masonry bricks due to the way energy is distributed by the material as it undergoes failure. This energy distribution is caused by energy being absorbed due to the composite nature of the material and is a direct result of the cellulose fibers in the mixture. To explain further, as the material begins to fail, the clay matrix starts to pull away from the cellulose fibers that—due to their flat helical shape—are well secured and absorb energy through two different methods. First, the helical fibers absorb energy as they uncurl and pull straight the same way a spring does, and second, they absorb energy by pulling out of the clay matrix.

The last method is known as fiber pullout strength. Fiber pullout strength is a very common energy absorption mechanism found in fiber-reinforced composites with weaker matrix interfaces similar to the matrix material observed herein [7,8].

During testing, it was noticed that the material exhibits a recovery zone in which, if the applied load is removed, the material will reform back to its cast shape. The recovery observation occurred on a visible scale due to the nature of the material's non-catastrophic failure method. A visible recovery was also noted post failure. A potential application of the material's deformation and recovery is as a buffer or indicator material to warn of system distress for brittle materials. This means that, on a visible inspection, if the ecobrick material is deformed, the load is within dangerous parameters of the other material and must be reduced.

The ecobrick material was also found to exhibit properties similar to wood, including its workability and ability to accept a wide range of surface treatments. The material can be used for many applications where wood or wood-based products are found. More testing is required on how readily this material accepts paints, stains, and shellacs, but early tests are promising.

Water absorption test samples were submerged for 24 h in a room-temperature water bath (70°F) of deionized water. After the 24-h duration, samples were removed from the water bath, towel-dried of exterior moisture, and their masses measured. Results of the water absorption test indicate all samples absorbed water, and the coating system failed to protect the samples from water ingress. Absorption is calculated to be a minimum of about 80 % among all samples. Numeric values of mass before and after submersion are shown in **Table 3**.

Analysis of Material Properties

Comparison of compressive strength of Ratio B, in **Table 1**, is on par with that of a second class brick. Unlike the second class brick, the brick being investigated does not require firing [2]. Firing bricks is the most energy-intensive part of brick production. It also produces the most carbon dioxide (CO_2) because gas kilns are often used. Therefore, removing this step makes this product much less costly and more environmentally friendly than the traditional brick firing methods.

Fig. 4 shows the compilation of the stress strain curves of the different test ratios assembled from the data. Ratio B ultimately failed in shear when the fibers separated from the clay matrix, as shown in **Fig. 5**. Understanding the mode of failure is

TABLE 3 Twenty-four-hour absorption test.

	Absorption (Weight %)		
Number of Coats	Ratio B	Ratio C	Ratio U
2	82.2	117	52.21
4	80.1	112	N/A
6	76.9	106	N/A

FIG. 4 Stress strain curves of six different ratio mixtures.

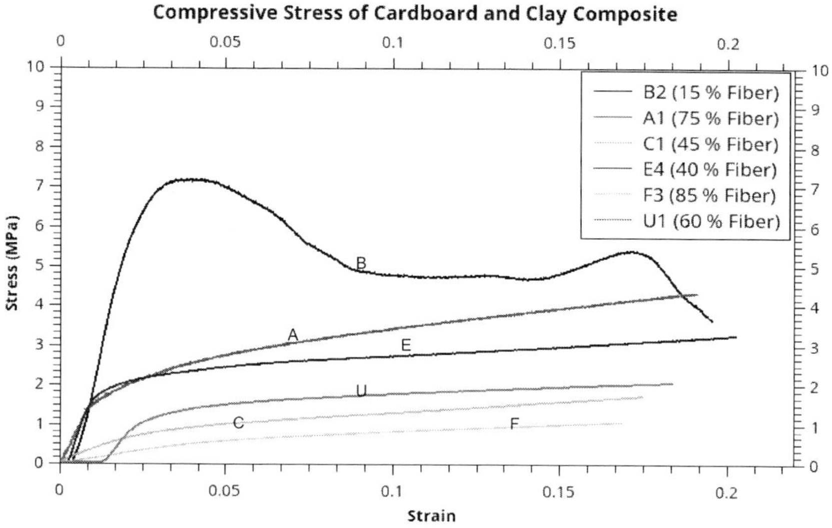

FIG. 5 Sample B2 prior to (a) and after the compression (b) test. Fibers were exposed as they were pulled out of the matrix during the shear failure (c).

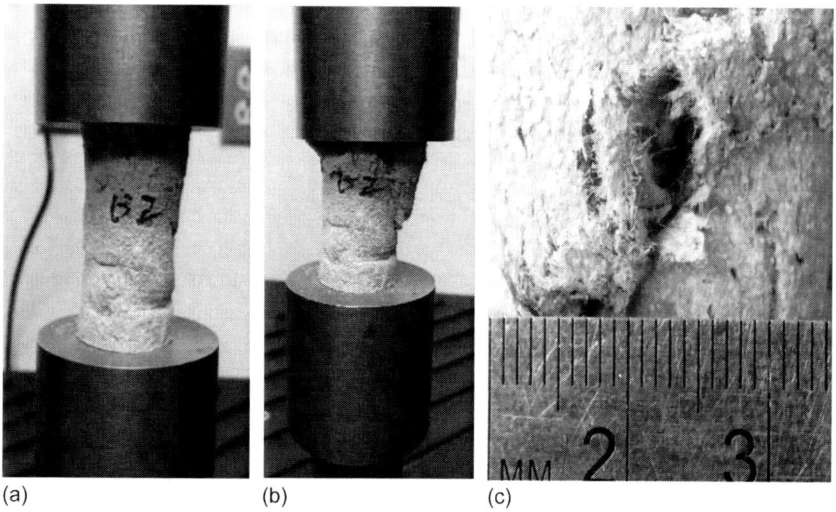

useful because a structure that is overloaded will show visible signs of distress while still maintaining structural integrity long enough for repairs to be done. Ratio B is also the only ratio to have an ultimate failure strength and not from a 2 % offset strain, as seen in **Table 1**. This is due to its lower void space and higher clay content that enables the fiber straightening and pullout strength to develop.

The modified Cox model is a prediction model used to estimate the combined properties of a composite material made up of a matrix material—in this case, bentonite clay and cellulose fibers [9,10]. This model is more accurate than the standard mixture rules for this application due to the use of the cellulose fiber because it takes the effective aspect ratio of the fibers into account when they are mixed in with the bentonite matrix. As this is a predictive model, it is designed to give approximate values for the composite mixture of clay and fiber similar to a standard mixture rule with more composite factors taken into account in the modeling, such as fiber pullout strength. The modified Cox model is used because it takes into account the features of the composite materials used, making it the most accurate predictive model available for the ecobrick.

Currently, the large amount of void space internally contained, particularly in the high fiber ratios, results in high porosity and low density. This leads to a breakdown of the initial model and results in some strengths being less than what is predicted by the modified Cox model because the model assumes minimal internal void space. This is particularly true of the higher fiber ratios such as Ratios A and D.

Another major problem that this material faces is that it readily absorbs water to the point of causing a loss of geometric shape. This was discovered after performing standard 24-h submersion tests on untreated samples. This test was then repeated on samples coated with increasing numbers of siloxane sealer coats, with the results shown in **Table 3**. The number of layers made a minimal difference in water absorption and all samples became waterlogged, gaining a minimum of approximately 80 % of their weight in water, as seen on **Table 3**. One possible solution is to look at what types of coatings are used for second class bricks because second class bricks have a similarly porous nature [11]. Despite the water absorption setback, the results of this project show great promise moving forward. As the human population continues to grow, demands for building materials will increase. Using waste material as a building material is both ecologically and economically responsible because it would cut down on human made CO_2 production from traditional clay brick. Postindustrial cellulose is also readily available as waste, and the clay required usually is available locally. Thus, both raw materials are low cost.

Future Directions

Future work will focus heavily on decreasing or eliminating void space in the material, as this will vastly increase the mechanical properties. A secondary task is testing more coatings and additives to better protect the material from water absorption. Investigations into the insulative properties (thermal and acoustic) and the fire resistance of the material are warranted. The R-value of the material needs to be evaluated, although it looks to be potentially very high due to the constituent cellulose fibers. This means that it has potential use as a load-bearing insulation. It is also easily worked using standard woodworking tools. Both of these qualities mean this material could have future applications in prefabricated or quick-rise housing or other similar structural applications. Further testing also should be done on its sound-insulating qualities to see how well this material blocks noise.

Attempts to further the structural aspect of this material and provide additional waterproofing with the use of siloxane is not performing as well as anticipated. Other coatings will need to be applied and tested, which should be easily accomplished as the material readily accepts surface coatings. These new coatings should be compared against both a control batch and the siloxane. For aesthetic purposes, looking into the addition of other additives such as carbon black will inherently modify the current material properties. As in the case of carbon black, increasing the ability of the material to absorb solar radiation thereby lowers the cost of heating any structure made of this material. Other potential additives that will need to be assessed for possible benefits include desiccant powders, antifungal agents, and potential odor neutralizers. All of these additives would provide different long-term benefits for specific situations and other potential benefits. Further testing will also be done to determine the exact amount of shrinkage present and the variations among ratios. This information would further help determine an ideal ratio as well as provide valuable information on the composite nature of the material and its potential limitations.

Another potential use for this material is as a nurse log for crops, particularly in arid areas or areas with poor soil nutrients. The material can hold its shape while absorbing a large amount of water, can be enriched with nutrients for plants (e.g., nitrates), and can be loaded with seeds and buried. This also has the advantage of fully recycling the waste material from which the brick is made.

Conclusions

Mechanical properties, compressive strength, and modulus of elasticity of a composite brick material is assessed by varying the ratio of the primary constituents (post-consumer cellulose and bentonite clay). The compressive strength tests performed show that, when properly densified, such as in Ratio B, the ecologically friendly brick matches or exceeds the compressive strength requirements for a second class fired brick—1,000 psi (6.9 MPa). The material's compressive strength ranges from approximately 338 psi (2.33 MPa) at low bentonite ratios (less than 50 % by mass) to approximately 1,015 psi (7.01 MPa) at higher bentonite ratios (greater than 60 % by mass) and has an elastic modulus of 340 MPa (49.3 KSI). Post-consumer cellulose is obtained from the waste stream of a cardboard recycling plant. The ecologically friendly brick saves energy by being air-dried, compared to kiln-fired, while simultaneously using post-consumer waste material as a major constituent. The primary raw material component used in the production of these units, cellulose, reduces landfill demand and the overall cost of production as compared with conventional brick. The material readily accepts surface coatings such as paint or sealers, and it can easily be tailored to specific purposes with the addition of small amounts of additional additives such as desiccants or colorant. A coating system was tested and evaluated for water resistance. Absorption is calculated to be a minimum of about 80 % among all samples tested. Currently, the largest hurdle is water absorption as this causes the material to swell and lose some of its structural integrity. Further research and testing is currently being performed to solve this problem, utilizing both surface treatments and internal additives. Testing also is being conducted on its insulative properties both acoustically and thermally.

ACKNOWLEDGMENTS

Support for this project was provided by the New Mexico Institute of Mining and Technology. Materials and expertise were provided by Frank Santoro of FS Enterprises. The authors especially thank the senior design students Amber Duran, Kevin Garber, Sawyer Gill, John M. O'Connell, and Alexandra Scheer.

References

[1] Ashori, A. and Nourbakhsh, A., "Characteristics of Wood-Fiber Plastic Composites Made of Recycled Materials," *Waste Manag.,* Vol. 29, No. 4, 2009, pp. 1291–1295.

[2] Raut, S., Ralegaonkar, R., and Mandavgane, S., "Development of Sustainable Construction Material Using Industrial and Agricultural Solid Waste: A Review of Waste-Create Bricks," *Constr. Build. Mater.*, Vol. 25, No. 10, 2011, pp. 4037–4042.

[3] Demir, I., Serhat Baspınar, M., and Orhan, M., "Utilization of Kraft Pulp Production Residues in Clay Brick Production," *Build. Environ.*, Vol. 40, No. 11, 2005, pp. 1533–1537.

[4] "Compressive/Crushing Strength of Bricks," *The Construction Civil*, 2017, https://web.archive.org/web/20170515045825/http://www.theconstructioncivil.org:80/compressive-crushing-strength-of-bricks/ (accessed April 12, 2017).

[5] Santoro, F. 2013. Products made from recycled cardboard. U.S. Patent 8,367,195, filed September 4, 2009, and issued February 5, 2013.

[6] ASTM C67-09, *Standard Test Methods for Sampling and Testing Brick and Structural Clay Tile*, ASTM International, West Conshohocken, PA, 2018, www.astm.org

[7] Callister, W. D., Jr., *Materials Science and Engineering: An Introduction*, 7th ed., Wiley, Hoboken, NJ, 2007.

[8] Shackelford, J. F., *Introduction to Materials Science for Engineers*, 6th ed., Pearson Education, Upper Saddle River, NJ, 2005.

[9] Gibson, R. F., *Principles of Composite Material Mechanics*, 4th ed., CRC Press, Taylor & Francis Group, Boca Raton, FL, 2016.

[10] Mallick, P. K., Ed., *Composites Engineering Handbook*, Taylor & Francis Group, Boca Raton, FL, 1997.

[11] "Classification of Bricks," *The Construction Civil*, 2017, https://web.archive.org/web/20170509031409/http://www.theconstructioncivil.org:80/classification-of-bricks/# (accessed April 12, 2017).

STP 1612, 2018 / available online at www.astm.org / doi: 10.1520/STP161220170151

John P. Sanders,[1] Nathaniel C. Huygen,[1]
and Stephen K. Smith[1]

Development and Calibration of Method for Thermal Property Measurement of Nonhomogeneous Masonry Materials

Citation

Sanders, J. P., Huygen, N. C., and Smith, S. K., "Development and Calibration of Method for Thermal Property Measurement of Nonhomogeneous Masonry Materials," *Masonry 2018, ASTM STP1612*, N. V. Krogstad and W. M. McGinley, Eds., ASTM International, West Conshohocken, PA, 2018, pp. 67–82, http://dx.doi.org/10.1520/STP161220170151[2]

ABSTRACT

Existing thermal conductivity or R-value data for masonry building materials are very limited and quite dated. Additionally, very little data on heat capacity are available. A convective heat transfer device was designed and built for the purpose of measuring the thermal performance of nonhomogeneous masonry materials that are not well suited for other testing methods. The device is similar to the hot box apparatus described in the standard ASTM C1363, *Standard Test Method for Thermal Performance of Building Materials and Envelope Assemblies by Means of a Hot Box Apparatus*, and it uses heat flux sensors to measure heat flows. This modified hot box apparatus is capable of measuring steady-state properties such as thermal conductivity or R-value. The modified hot box can also measure non-steady-state heat capacity that is traditionally measured by laser/light flash apparatuses. The heat capacity measurements can be used to calculate the thermal diffusivity when the density and thermal conductivity are

Manuscript received September 7, 2017; accepted for publication December 6, 2017.
[1]National Brick Research Center–Clemson University, 100 Clemson Research Blvd., Anderson, SC 29625, USA
J. P. S. http://orcid.org/0000-0003-1655-8319
[2]ASTM International Symposium on *Masonry 2018* on June 26, 2018 in San Diego, CA, USA.

known. The modified hot box instrumentation was calibrated to acquire data on building materials such as brick, concrete, and other nonhomogeneous materials that were large in size and mass.

Keywords

R-value, masonry materials, heat capacity, thermal conductivity

Introduction

Due to their large size and nonhomogeneous nature, bricks, mortars, concrete, and other building materials benefit from bulk sample methodologies for the measurement of thermal properties. Thermal conductivity testing hardware such as the Netzsch HFM 436 Lambda (NETZSCH, Burlington, MA), which is based on ASTM C518, *Standard Test Method for Steady-State Thermal Transmission Properties by Means of the Heat Flow Meter Apparatus* [1], can accept large specimens with dimensions of 30 cm by 30 cm (12 in. by 12 in.) and with a thickness between 0.05 cm (0.02 in.) and 10 cm (4 in.) [2]. Although the apparatus can take large samples, the range of thermal conductivities that can be measured is limited to between 0.005 W/m × K (0.035 BTU/in × hr × ft^2 × F) and 0.5 W/m × K (3.47 BTU/in × hr × ft^2 × F) [2], which lies outside the range of most bricks, mortars, and concrete. Most of these materials have thermal conductivity greater than 1 W/m × K (6.9 BTU/in × hr × ft^2 × F) [3]. Laser and light flash techniques are well suited for characterizing homogenous materials. The technique was first described by Parker, Jenkins, Butler, and Abbott in 1961 [4], where samples were pulsed with a light of a known amount of energy, and the responding temperature response on the rear of the sample was recorded. Samples were coated in a highly absorptive coating that enhanced the energy absorption into the specimen, which was critical for making assumptions as to the amount of energy actually input into the sample [4,5]. Laser and light flash devices primarily give information on the thermal diffusivity of the sample, with thermal conductivity and heat capacity being extrapolated from the temperature profile curve [4]. With these devices, the main limitation is sample size. Samples are generally on the order of 0.1 cm (0.04 in.) thick with a surface area less than 5 cm^2 (0.78 in^2) [4]. Another device to measure transient thermal properties is the transient plane sensor device first described by Gustafsson in 1991 [6], where he used a thin wire heater that acted as both a heater and sensor for probing the thermal properties of a wide array of materials. A transient plane sensor device is best used on materials with thermal conductivity between 0.1 W/m × K (0.69 BTU/in × hr × ft^2 × F) and 100 W/m × K (693 BTU/in × hr × ft^2 × F) [6]. These limits are due to the long measurement time required for highly insulative materials and to thermal contact resistance in highly conductive materials [6]. Sample sizes for this technique are in the range of 2 cm (0.79 in.) thick and 5 cm^2 (0.78 in^2). Nonhomogeneous materials can change the results of this method when microstructural features begin to approach the size of the sensor being used. Depending on sensor placement, acquiring different results can give a semi-quantitative view of the degree of homogeneity of the specimen [7]. Bricks, mortar, and concrete all have varying degrees of homogeneity and larger microstructural features and thus benefit from having a larger sample to analyze.

Heat flux transducers (HFTs) provide a unique method of capturing data on the energy density propagating through a material. HFTs work by using thermocouples wired in series (a thermopile), embedded in a substrate, that produce changes in voltage as a function of applied temperature gradient. With proper calibration, HFTs enable the direct measurement of the heat flux through a sample without relying on estimations of heat flow based upon energy input and energy loss as is traditionally done in a calibrated hot box [8]. The ability to directly measure heat flux allows for evaluation of dynamic thermal performance and calculation of heat capacity and thermal diffusivity. The use of HFTs for measuring thermal properties under steady-state conditions is described in ASTM C518 [9], EN 1934 [10], and ISO 8301 [11]. HFTs have been used to measure the heat flux under dynamic loading conditions [12–14] but have not been used to directly measure thermal properties such as heat capacity and thermal diffusivity, insofar as we are aware.

Experimental

MODIFIED HOT BOX INTRODUCTION

The modified hot box (MHB) was structurally based on a hot box device, as specified in ASTM C1363, *Standard Test Method for Thermal Performance of Building Materials and Envelope Assemblies by Means of a Hot Box Apparatus* [8]. The MHB used convective heat transfer to impose thermal gradients across a sample. The MHB is shown in **Fig. 1** and **Fig. 2**. Peltier thermoelectric modules were used to heat

FIG. 1 Modified hot box apparatus.

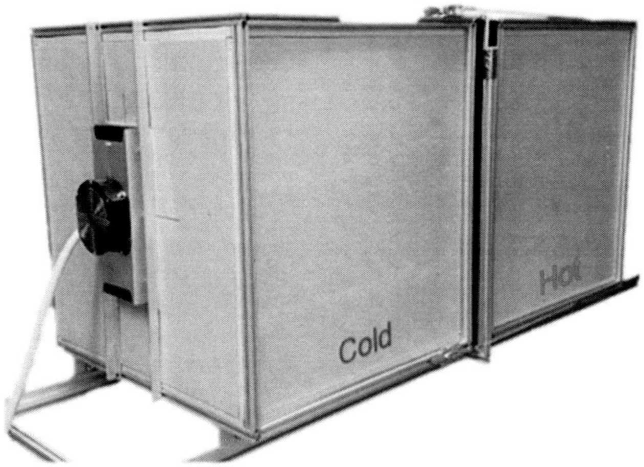

FIG. 2 Cross section of modified hot box.

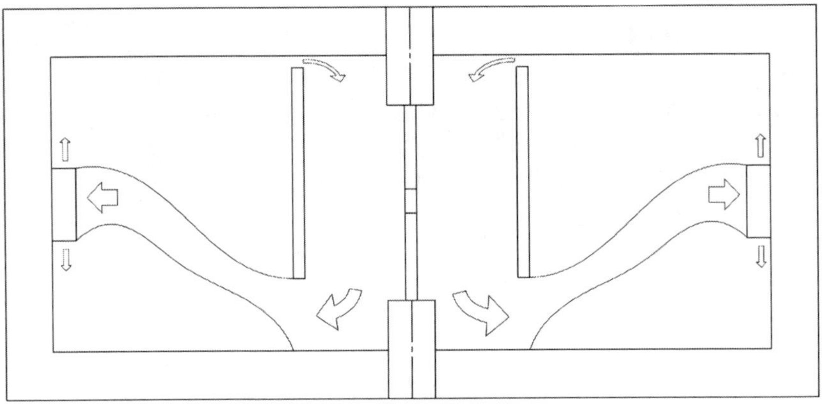

or cool the air contained in the insulated chambers of the MHB. The chambers were highly insulated with a layering of extruded polystyrene (XPS) board that was 10 cm (4 in.) thick.

Heat flow was measured using HFTs that were 5 cm by 5 cm by 0.3 cm (2 in. by 2 in. by 1/8 in.) and were made of polyimide. Sensitivity coefficients for the HFTs were determined according to ASTM C1130, *Standard Practice for Calibration of Thin Heat Flux Transducers* [15]. Sample dimensions were 5 cm by 5 cm (2 in. by 2 in.) so that they were completely covered by the HFTs. The sample and HFTs were then placed within the XPS surround with only the HFTs exposed to the air. This configuration is regarded as a guarded setup. The guarded setup is important to help ensure one-dimensional heat flow through the sample and HFTs. Additionally, it serves to guarantee that the measured heat flow is actually going through the sample and not around it. An example of this guarded setup is shown in **Fig. 3**.

Temperature sensing was performed using surface-mounted 1,000-ohm platinum resistance thermometer detectors (RTDs). Measurement of air temperature was done using a 10,000-ohm negative temperature coefficient thermistor. This thermistor served as the input for the PID (proportional—integral—derivative) controllers that control the thermoelectric modules in order to maintain temperature set points. The sample with attached RTDs and HFTs placed within an XPS frame is shown in **Fig. 4**.

CALIBRATION OF THE MHB

The MHB was calibrated for testing thermal conductivity and heat capacity using several common materials. These materials had thermal properties that were

FIG. 3 Sample in guarded configuration with HFTs.

FIG. 4 Sample, HFTs, and RTDs within XPS frame.

representative of the range of materials expected to be tested in the MHB. The calibration materials used are shown in Table 1.

Calibration of the MHB was broken down into two separate procedures—one for steady-state measurements and the other for transient measurements. In order

TABLE 1 Properties of calibration standards.

Sample	Density [kg/m^3] (lb/ft^3)	Thickness [mm] (in.)	Conductivity [W/mK] (BTU * ft/hr * ft^2 * F)	Heat Capacity [J/kgK] (BTU/lb * F)
Sheetrock	637 (39.8)	12.68 (0.5)	0.16 (0.09) [16]	1,090 (0.26) [16]
Marble	2,680 (167.3)	9.93 (0.39)	2.62 (1.51) [17]	896 (0.214) [17]
HDPE-3/8 in.	960 (59.9)	9.55 (0.38)	0.42 (0.24) [18]	1,838 (0.439)[a]
HDPE-1 in.	966 (60.3)	25.21 (0.99)	0.48 (0.28) [18]	1,838 (0.439)[a]
NIST SRM 1453d	41 (2.6)	13.25 (0.52)	0.033 (0.019) [19]	1,470 (0.351) [20]
Acrylic	1,174 (73.3)	5.1 (0.2)	0.14 (0.08) [21]	1,460 (0.349) [22,23]
XPS-1 in.	34 (2.1)	23.43 (0.92)	0.033 (0.019) [24]	1,280 (0.305) [10]

Note: [a]Heat capacity of HDPE was determined using calorimetry.

to be able to measure both transient and steady-state data, the system was initially brought to thermal equilibrium with both chambers having an air temperature of 9°C (48.2°F). Once the system reached equilibrium, the air temperature on one side was stepped to 39°C (102.2°F). Steady-state data were collected once the system had reached a constant heat flux. Typical time to reach steady state was strongly dependent on the thermal mass of the samples. For example, the 1-in. XPS took 1.3 h to reach thermal equilibrium, whereas the 1-in. HDPE took 2.0 h. Steady-state data were collected for a period of time equaling 25 % of the amount of time it took the system to reach equilibrium. Total analysis time—the time required for initial equilibration, subsequent application of a thermal gradient, and return to the original initial condition—was 7.8 h for 1-in. XPS and 14.7 h for 1-in. HDPE. This cycle is presented in **Fig. 5**. The more thermally massive samples took significantly longer to test.

SAMPLE PREPARATION AND MEASUREMENT CONSIDERATIONS

When performing thermal measurements using HFTs, care must be taken that the HFTs are in intimate contact with the surface of the sample. Since the HFTs have a smooth, flat surface, the material being measured must either be semi-deformable or also have a smooth, flat surface in order to have good contact with the HFTs. This is accomplished by surface grinding the samples, which also ensures uniform thickness across the sample. Surface grinding treatments are recommended by ASTM C177-13, *Standard Test Method for Steady-State Heat Flux Measurements and Thermal Transmission Properties by Means of the Guarded-Hot-Plate Apparatus* [1], and ASTM C518 [9]. When surface grinding was not feasible, a thin layer of Dow Corning 340 Thermal Heat Sink Compound (Dow Corning, Midland, MI) was applied, which improves contact between the sample and the HFTs. The thermal conductivity of this heat sink compound was comparable to the HFTs themselves, and keeping this layer thin ensured measurement accuracy.

Thermal contact resistance is a function of several parameters such as the temperature, pressure and mechanical properties of the test, and the sensor materials

FIG. 5 Temperatures during a typical measurement cycle.

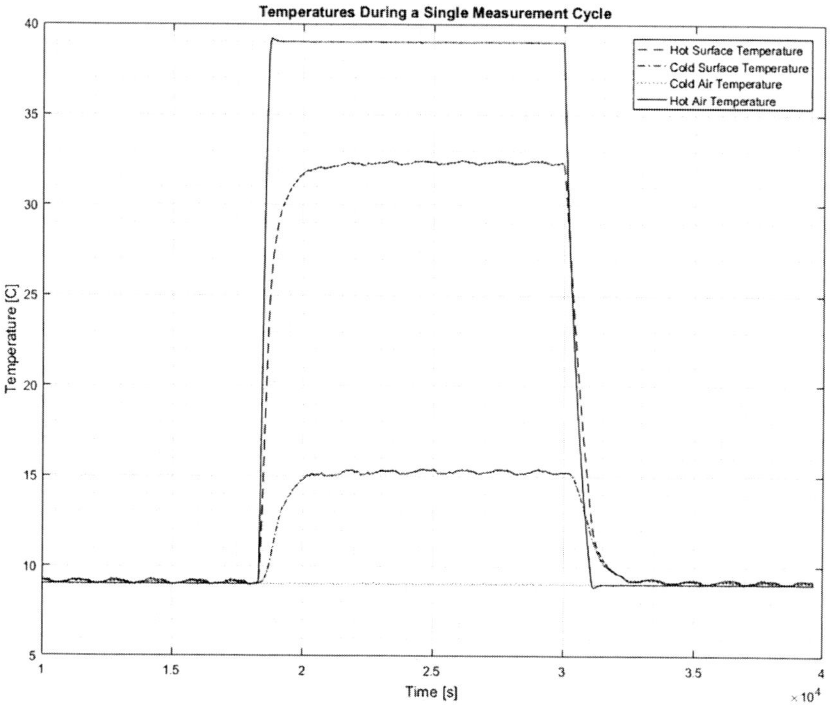

playing roles in how heat flow interacts between two boundaries [25–27]. The Netzsch HFM 436 Lambda Heat Flux Meter can reportedly measure thermal conductivity within 5 % when the reference material is between 50 % and 200 % of the thermal resistance of the material being tested [2]. Similar observations of the limits of calibration were taken into account by Tleoubaev and Brzezinski, who included the effect of thermal contact resistance in his calibration [28].

HEAT CAPACITY MEASUREMENTS BY SIMPLE CALORIMETER

Heat capacity measurements were performed in a simple calorimeter on the calibration samples. Samples were equilibrated for at least 24 h in a laboratory environment where air temperature was maintained at 23°C (73.4°F). The mass of the samples was recorded using a calibrated Sartorius CP224S Mass Balance (Sartorius, Goettingen, Germany) prior to being heated in the oven. Samples were heated in a laboratory oven for at least 1 h. The laboratory oven temperature had been verified by the calibrated Omega CL25 Calibrator-Thermometer (Omega, Norwalk, CT) using a T-type thermocouple that sampled air temperature slightly above the samples. The simple calorimeter used water as the fluid bath, which came from a source that had been equilibrated with the laboratory environment overnight.

When the samples were assumed to be at thermal steady state, they were transferred into the calorimeter fluid bath, where temperature was monitored using a calibrated Omega CL25 Calibrator-Thermometer, sensed by a T-type thermocouple. The calorimeter was placed on a VWR Analog Shaker (VWR, Brooklyn, NY) to gently stir the fluid bath during the measurement.

Results and Discussion

CALIBRATION OF THE MHB (STEADY STATE)

Comparing raw data from the MHB, there were large deviations in the measured thermal conductivity relative to those reported in literature that were believed to be due to the presence of thermal contact resistance between the sample and HFTs. A plot presented in **Fig. 6** shows the measured thermal conductivity versus the true (literature) thermal conductivity. Significant deviations from expected values were observed.

In order to try to account for this disparity between measured conductivity and true conductivity, the sample's thermal resistance was evaluated. The measured R-value versus literature R-value is presented in **Fig. 7**. A linear regression was

FIG. 6 Thermal conductivity, measured versus literature.

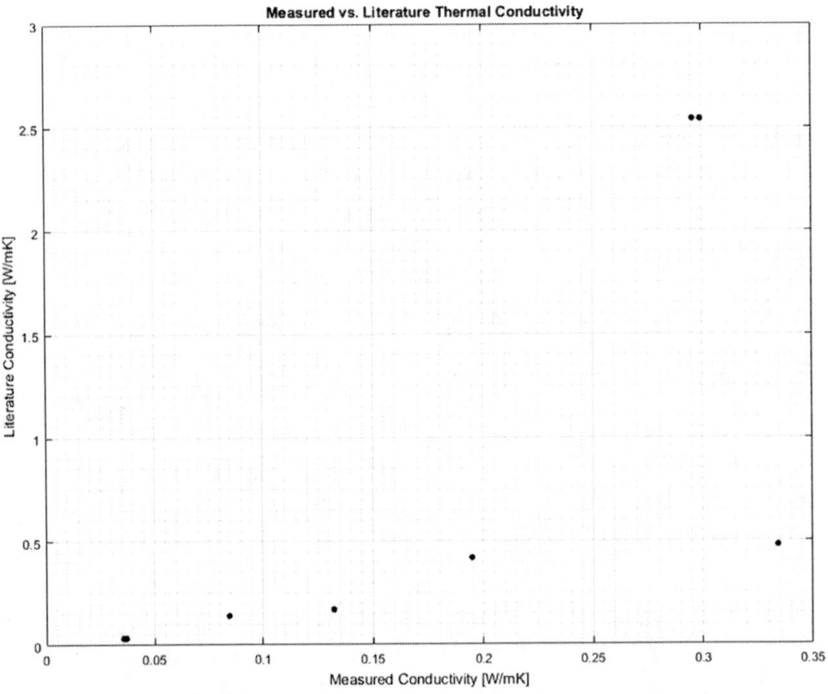

FIG. 7 Measured versus literature thermal resistance.

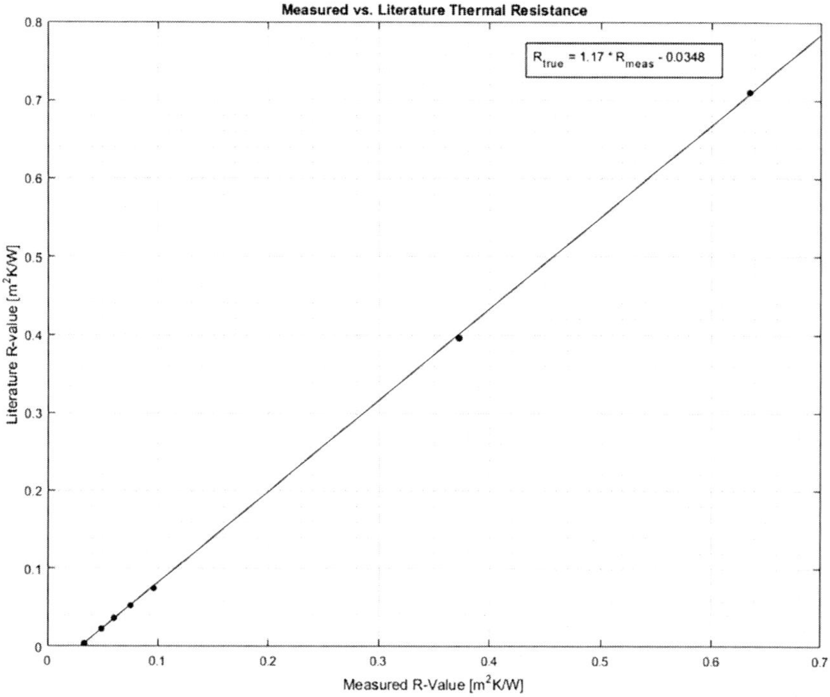

performed on the data such that the relative error of the corrected conductivity was minimized. This was necessary due to obvious calculation errors encountered with highly conductive samples. This was because the higher conductivity samples had very small R-values and the calculated conductivity was very sensitive to the calibration coefficients. This was also partly due to the large range of R-values being calibrated (0.0038–0.71 m^2K/W; 0.022–4.0 $ft^2 \times h \times F/BTU$). Simply put, it was a weighting issue.

The measured R-value and the corrected R-value are given by Eq 1:

$$R_{meas} = \frac{q''}{\Delta T}$$

$$R_{true} = 1.17 * R_{meas} - 0.0348 \tag{1}$$

where q'' is the heat flux recorded by the HFTs, and $\Delta T$is the surface temperature difference recorded by the platinum RTDs.

The offset in the calibration equation (Eq 1) is the manifestation of the thermal contact resistance between RTDs and the sample surface as well as the R-value of the HFTs themselves. The measured R-value is the sum of the R-values of the sample, both HFTs, and the contact resistance between HFTs and the sample. This contact resistance is caused by the interface between two materials. It can be minimized by having better contact between the two surfaces, but it cannot be eliminated entirely.

Thermal conductivity can be calculated using the corrected thermal resistance and sample thickness. The corrected thermal conductivity is given by Eq 2:

$$k_{corr} = \frac{l}{R_{corr}} \tag{2}$$

where l is the thickness of the sample, and R_{corr} is the corrected thermal resistance as given by Eq 2.

The corrected thermal conductivity plotted against the literature conductivity is shown in **Fig. 8**. The spread of measurements at high conductivity was due to how close the measured R-value is to the offset in the calibration equation. The measurement standard deviation on a single run was typically in the range of 0.5 %. For example, the literature value for the R-value of the marble was given as 0.0039 m^2K/W (0.022 $ft^2 \times h \times F/BTU$). The measured R-value for both data points for marble in **Fig. 8** was 0.0331 m^2K/W (0.188 $ft^2 \times h \times F/BTU$) and

FIG. 8 Thermal conductivity, corrected versus literature.

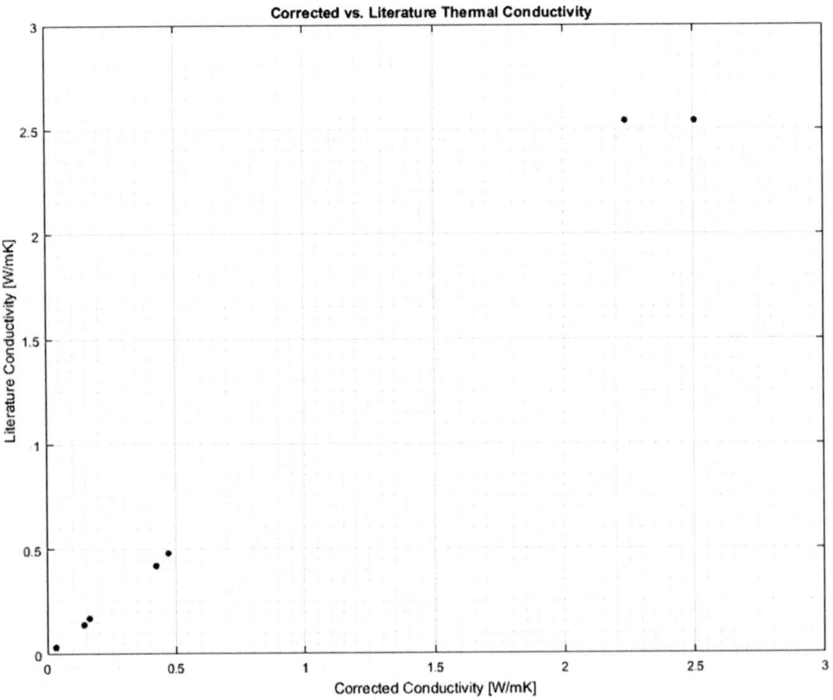

0.0336 m^2K/W (0.191 $ft^2 \times h \times F/BTU$). For this particular sample, the thermal contact resistance and R-value of the HFTs represented almost 90 % of the measured value. If the measured standard deviation is 0.5 %, and the sample only accounts for 10 % of the total R-value, then the standard deviation on the sample's R-value is not 0.5 % but instead is 5 %. For high conductivity materials, obtaining more accurate results will require testing thicker samples. The larger the R-value measured, the higher the confidence the corrected conductivity will be accurate.

CALIBRATION OF THE MHB (TRANSIENT STATE)

One of the purposes of the MHB was to be able to measure not only the thermal conductivity of materials but also their heat capacity. The heat fluxes and temperatures during a single temperature step are shown in Fig. 9. Looking at the heat flux versus time curve, there is an area between the two heat flux curves that represents the difference between the energy that enters the sample and the energy that leaves the sample. This energy is responsible for the temperature change of the sample during step response. Intuitively, this is related to the heat capacity of the sample.

FIG. 9 Heat flux versus time and temperature versus time during a single step response.

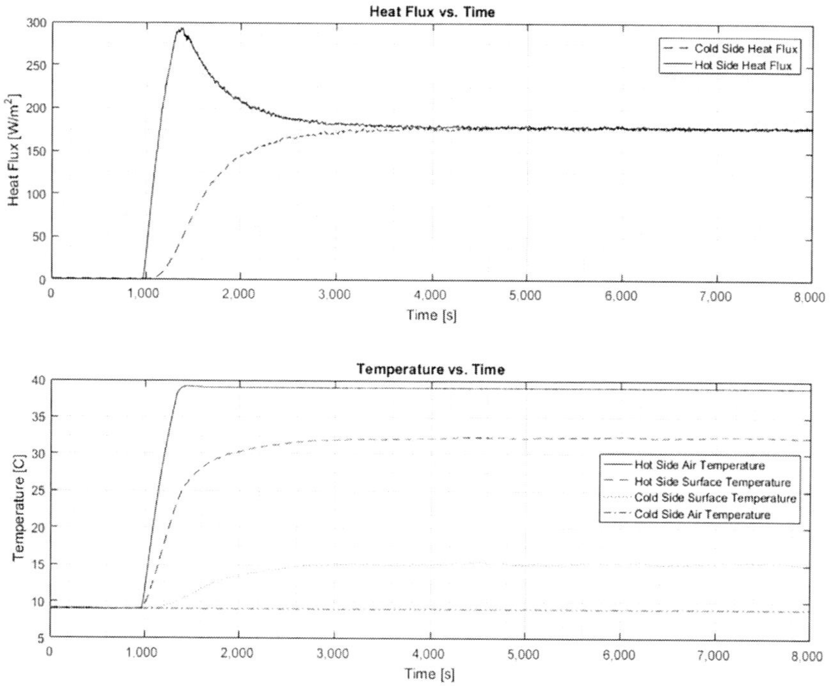

The justification for this lies in the analysis of the one-dimensional solution of the heat equation with convection boundary conditions. Given a material with thickness l, conductivity k, density ρ, heat capacity C_P, air temperatures of T_∞ and T_0, film coefficient h, and initial temperature T_∞, the problem can be stated by Eq 3:

$$\frac{\partial T}{\partial t} = \frac{k}{\rho C_P}\frac{\partial^2 T}{\partial x^2}$$
$$-k\frac{\partial T}{\partial x}\bigg|_{x=0} + h(T(0,t) - T_0) = 0$$
$$k\frac{\partial T}{\partial x}\bigg|_{x=l} + h(T(l,t) - T_\infty) = 0 \tag{3}$$

The solution to the problem stated in Eq 3 is given by Eq 4 [29]:

$$T(x,t) = \frac{\left((T_\infty - T_0)\left(\frac{h}{k}\right)x + \left(\left(\frac{h}{k}\right)l + 1\right)T_0 + T_\infty\right)}{\left(\frac{h}{k}\right)l + 2}$$
$$+2\sum_{n=1}^{\infty}\frac{X(x)\hat{T}(t)}{\left(\mu_n^2 + \left(\frac{h}{k}\right)^2\right)l + 2\left(\frac{h}{k}\right)}\int_0^l \frac{\left((T_\infty - T_0)\left(1 + \left(\frac{h}{k}\right)l - \left(\frac{h}{k}\right)x\right)\right)}{\left(\frac{h}{k}\right)l + 2}X(x)dx$$
$$X(x) = \mu_n\cos(\mu_n x) + \left(\frac{h}{k}\right)\sin(\mu_n x)$$
$$\hat{T}(t) = \exp\left(-\mu_n^2\left(\frac{k}{\rho C_P}\right)t\right) \tag{4}$$

And μ_n is given by the solution to the following transcendental equation (Eq 5):

$$\tan(\mu_n l) = \frac{2\left(\frac{h}{k}\right)\mu_n}{\mu_n^2 - \left(\frac{h}{k}\right)^2} \tag{5}$$

In order to analyze the energy storage represented by the area between the two heat flux curves, Eq 6 was performed on the previous solution.

$$\int_0^\infty -k\frac{\partial T}{\partial x}\bigg|_{x=0} + k\frac{\partial T}{\partial x}\bigg|_{x=l} dt \tag{6}$$

Evaluating this numerically for several different values of each physical parameter yielded the following equivalent expression (Eq 7):

$$\frac{\rho C_P l(T_\infty - T_0)}{2} \tag{7}$$

From several of the samples that have been tested, the surface temperature response was not entirely symmetric with regard to the applied air temperature. When a mean air temperature of 24°C (75.2°F) was applied, the same mean temperature was not observed with regard to the surface temperature. For example, a mean temperature of 23.7°C (74.7°F) was typically observed; however, this varies slightly from sample to sample. Taking this into consideration, instead of using $T_\infty - T_0$ in Eq 7, it is instead replaced with $\bar{T}_{new} - \bar{T}_{old}$, where $\bar{T}_{old}$ is the mean temperature before the step in temperature, and $\bar{T}_{new}$ is the mean temperature after the temperature step.

One final consideration that must be taken into account before heat capacity can be accurately measured is the heat storage of the HFTs. **Fig. 10** shows the heat capacity calibration curve that was devised using the previous methods. Note that what is being calibrated is the measured area of the heat flux difference curve divided by $\frac{1}{2}(\bar{T}_{new} - \bar{T}_{old})$, which represents the energy stored per degree change for the entire assembly (sample + HFTs). The extra energy stored by the HFTs should manifest itself as an offset in the calibration equation.

Deviation from expected linear behavior occurs in samples with a very large heat storage. The measured energy was smaller than expected and was attributed to heat loss through the surround due to the long time required for the sample to reach equilibrium. The final heat capacity was determined by taking the corrected

FIG. 10 Measured versus expected energy storage.

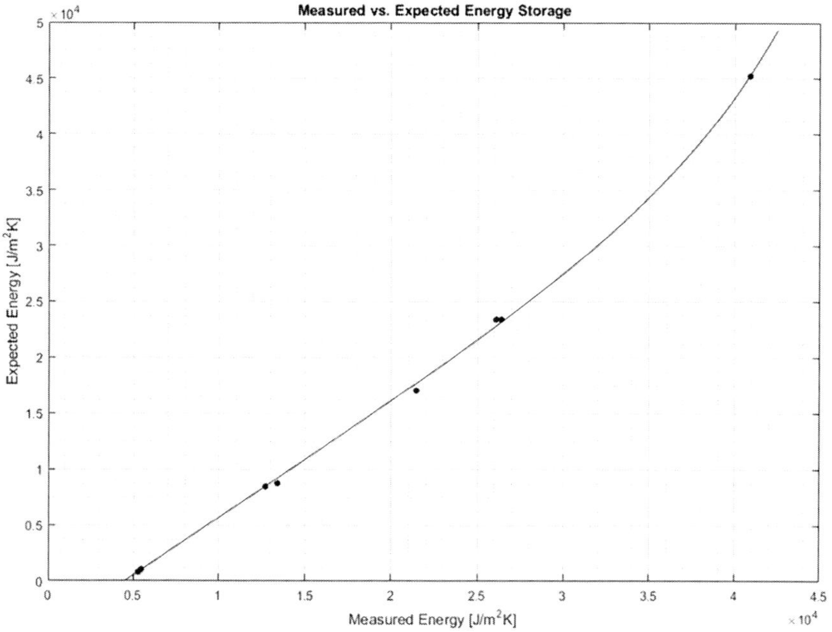

TABLE 2 Thermal conductivity and heat capacity results for calibration samples..

Sample	Literature Conductivity [W/mK] (BTU*ft/ hr * ft^2 * F)	Corrected Conductivity [W/mK] (BTU*ft/ hr * ft^2 * F)	% Difference	Literature Heat Capacity [J/kgK] (BTU/lb * F)	Corrected Heat Capacity [J/kgK] (BTU/lb * F)	% Difference
Gypsum	0.16 (0.09)	0.163 (0.09)	1 %	1,090 (0.26)	1,054 (0.252)	3 %
Marble	2.62 (1.51)	2.47 (1.43)	6 %	896 (0.214)	864 (0.206)	4 %
HDPE 3/8 in.	0.42 (0.24)	0.424 (0.24)	1 %	1,838 (0.439)	1,934 (0.462)	5 %
1453d	0.033 (0.019)	0.033 (0.019)	0 %	1,465 (0.35)	1,501 (0.359)	2 %
Acrylic	0.14 (0.08)	0.142 (0.08)	1 %	1,460 (0.349)	1,543 (0.369)	6 %
HDPE 1 in.	0.47 (0.27)	0.47 (0.27)	0 %	1,838 (0.439)	1,840 (0.439)	0 %
XPS 1 in.	0.033 (0.019)	0.033 (0.019)	0 %	1,280 (0.306)	1,230 (0.294)	4 %

energy storage and dividing by the sample density and thickness. The sample heat capacity is given by Eq 8:

$$C_P = \frac{(1.03Area - 4614)(1 + 0.0008\exp(1.35 * 10^{-4}Area))}{\rho l} \tag{8}$$

THERMAL PROPERTY RESULTS FOR CALIBRATION SPECIMENS

The final results of the calibration for the thermal conductivity and heat capacity are given in **Table 2**. The mean deviation across all calibration samples for the thermal conductivity was 1 % and heat capacity was 3 %.

Conclusions

In this research, a new convective heat transfer device was built for testing building materials to measure thermal conductivity and heat capacity. It was found that the modified hot box was capable of acquiring this data using heat flux transducers and resistance thermometer detectors.

The modified hot box apparatus was developed to acquire thermal conductivity data within 1 % of literature values and heat capacity to within 3 % of literature values for calibration specimens on average. As a first of its kind of system, the modified hot box provides reliable and acceptable results.

References

[1] ASTM C177-13, *Standard Test Method for Steady-State Heat Flux Measurements and Thermal Transmission Properties by Means of the Guarded-Hot-Plate Apparatus*, ASTM International, West Conshohocken, PA, 2013, www.astm.org

[2] NETZSCH, *Instrument Manual—Netzsch HFM 436 Lambda*, NETZSCH-Geratebaue GmbH, Selb, Germany, 2011.

[3] Valore, R., Tuluca, A., and Caputo, A., "Assessment of the Thermal and Physical Properties of Masonry Block Products," Report No. ORNL/Sub/86-22020/1, Oak Ridge National Laboratory Metal and Ceramics Division, Oak Ridge, TN, 1988.

[4] Parker, W. J., Jenkins, R. J., Butler, C. P., and Abbott, G. L., "Flash Method of Determining Thermal Diffusivity, Heat Capacity, and Thermal Conductivity," *J. Appl. Phys.,* Vol. 32, No. 9, 1961, pp. 1679–1684, http://dx.doi.org/10.1063/1.1728417

[5] Rémy, B., Maillet, D., and André, S., "Laser-Flash Diffusivity Measurement of Diamond Films," *Int. J. Thermophys.,* Vol. 19, No. 3, 1998, pp. 951–959, http://dx.doi.org/10.1023/A:1022663611392

[6] Gustafsson, S. E., "Transient Plane Source Techniques for Thermal Conductivity and Thermal Diffusivity Measurements of Solid Materials," *Rev. Sci. Instrum.*, Vol. 62, No. 3, 1991, pp. 797–804, http://dx.doi.org/10.1063/1.1142087

[7] Sizov, A., Cederkrantz, D., Salmi, L., Rosén, A., Jacobson, L., Gustafsson, S. E., and Gustavsson, M., "Thermal Conductivity versus Depth Profiling of Inhomogeneous Materials Using the Hot Disc Technique," *Rev. Sci. Instrum.*, Vol. 87, No. 7, 2016, 074901, http://dx.doi.org/10.1063/1.4954972

[8] ASTM C1363-11, *Standard Test Method for Thermal Performance of Building Materials and Envelope Assemblies by Means of a Hot Box Apparatus*, ASTM International, West Conshohocken, PA, 2011, www.astm.org

[9] ASTM C518-17, *Standard Test Method for Steady-State Thermal Transmission Properties by Means of the Heat Flow Meter Apparatus*, ASTM International, West Conshohocken, PA, 2017, www.astm.org

[10] EN 1934:1998, *Thermal Performance of Buildings - Determination of Thermal Resistance by Hot Box Method Using Heat Flow Meter—Masonry,* European Committee for Standardization, Brussels, Belgium, 1998.

[11] ISO 8301:1991(E), *Thermal Insulation—Determination of Steady-State Thermal Resistance and Related Properties—Heat Flow Meter Apparatus*, International Organization for Standardization, Geneva, Switzerland, 1991.

[12] Sala, J., Urresti, A., Martín, K., Fores, I., and Apaolaza, A., "Effect of Thermal Mass on the Thermal Performance of Various Australian Residential Construction Systems," *Energ. Buildings,* Vol. 40, No. 4, 2008, pp. 459–465.

[13] Gori, V., Marincioni, V., Biddulph, P., and Elwell, C. A., "Inferring the Thermal Resistance and Effective Thermal Mass Distribution of a Wall from in Situ Measurements to Characterise Heat Transfer at Both the Interior and Exterior Surfaces," *Energ. Buildings*, Vol. 135, 2017, pp. 398–409, http://dx.doi.org/10.1016/j.enbuild.2016.10.043

[14] Biddulph, P., Gori, V., Elwell, C. A., Scott, C., Rye, C., Lowe, R., and Oreszczyn, T., "Inferring the Thermal Resistance and Effective Thermal Mass of a Wall Using Frequent Temperature and Heat Flux Measurements," *Energ. Buildings*, Vol. 78, 2014, pp. 10–16, http://dx.doi.org/10.1016/j.enbuild.2014.04.004

[15] ASTM C1130-17, *Standard Practice for Calibrating Thin Heat Flux Transducers*, ASTM International, West Conshohocken, PA, 2012, www.astm.org

[16] Gypsum Association, *Gypsum Board Typical Mechanical and Physical Properties,* GA-235-10, Gypsum Association, Hyattsville, MD, 2010, https://web.archive.org/web/20170517081055/http://www.americangypsum.com/sites/default/files/documents/GA-235%20Gypsum%20Board%20Typical%20Mechanical%20and%20Physical%20Properties.pdf (accessed July 17, 2017).

[17] Peirce, B. O. and Willson, R. W., "On the Thermal Diffusivities of Different Kinds of Marble," *P. Am. Acad. Arts Sci.*, Vol. 36, No. 2, 1900, pp. 13–16.

[18] "High-Density Polyethylene," https://web.archive.org/web/20170324003649/http://polymerdatabase.com/Commercial%20Polymers/HDPE.html (accessed March 24, 2017).

[19] Zarr, R. R. and Pintar, A. L. "Standard Reference Materials: SRM 1453, Expanded Polystyrene Board, for Thermal Conductivity from 281 K to 313 K," National Institute of Standards and Technology, U.S. Department of Commerce, Gaithersburg, MD, 2012, http://dx.doi.org/10.6028/nist.sp.260-175

[20] American Society of Heating, Refrigerating and Air-Conditioning Engineers, *2013 ASHRAE Handbook: Fundamentals*, ASHRAE, Atlanta, GA, 2013.

[21] Plaskolite, "OPTIX Properties: Physical, Chemical, Mechanical, and Thermal Properties of OPTIX Acrylic Sheet, Plaskolite, Columbus, OH, 2017, https://web.archive.org/web/20170709223025/http://www.plaskolite.com/Fabrication/Acrylic/Optix (accessed July 9, 2017).

[22] Professional Plastics, "Thermal Properties of Plastic Materials," Professional Plastics, Fullerton, CA, 2017, https://web.archive.org/web/20170726113112/http://www.professionalplastics.com/professionalplastics/ThermalPropertiesofPlasticMaterials.pdf (accessed July 26, 2017).

[23] Yucel, K. T., Basyigit, C., and Ozel, G., "Thermal Insulation Properties of Expanded Polystyrene as Construction and Insulating Materials," Suleyman Demirel University, Isparta, Turkey, 2003, https://web.archive.org/web/20160324174400/zenonpanel.com.mk/al/wp-content/uploads/2009/06/Thermal-Insulation-properties.pdf (accessed March 24, 2017).

[24] Al-Ajlan, S. A., "Measurements of Thermal Properties of Insulation Materials by Using Transient Plane Source Technique," *Appl. Therm. Eng.*, Vol. 26, Nos. 17–18, 2006, pp. 2184–2191.

[25] Mantelli, M., Yovanovich, M., and Gilmore, D., "Thermal Contact Resistance," *Spacecraft Thermal Control Handbook*, Vol. 1, D. G. Gilmore, Ed., American Institute of Aeronautics and Astronautics, Inc., Reston, VA, 2002, pp. 599–638.

[26] Mikic, B. B. and Rohsenow, W. M., "Thermal Contact Resistance," Report No. DSR 74542-41, MIT Department of Mechanical Engineering, Cambridge, MA, 1966.

[27] Salgon, J. J., Robbe-Valloire, F., Blouet, J., and Bransier, J., "A Mechanical and Geometrical Approach to Thermal Contact Resistance," *Int. J. Heat Mass Tran.*, Vol. 40, No. 5, 1997, pp. 1121–1129, http://dx.doi.org.libproxy.clemson.edu/10.1016/0017-9310(96)00130-5

[28] Tleoubaev, A. and Brzezinski, A., "Errors of the Heat Flow Meter Method Caused by Thermal Contact Resistance," *Proceedings of the 29th International Thermal Conductivity Conference and the 17th International Thermal Expansion Symposium,* June 24–27, 2007, Birmingham, AL, J. R. Koenig and H. Ban, Eds., DEStech Publishing, Inc., Lancaster, PA, 2008, pp. 465–473.

[29] Carslaw, H. S. and Jaeger, J. C., *Conduction of Heat in Solids*, Clarendon Press, New York, 2011.

STP 1612, 2018 / available online at www.astm.org / doi: 10.1520/STP161220170173

Nathaniel Huygen[1] and John P. Sanders[1]

Dynamic Thermal Performance Measurements of Residential Wall Systems

Citation

Huygen, N. and Sanders, J. P., "Dynamic Thermal Performance Measurements of Residential Wall Systems," *Masonry 2018, ASTM STP1612*, N. V. Krogstad and W. M. McGinley, Eds., ASTM International, West Conshohocken, PA, 2018, pp. 83–106, http://dx.doi.org/10.1520/STP161220170173[2]

ABSTRACT

The use of a hot box apparatus allows for thermal performance testing of a specimen in both static and dynamic regimes. In the static (steady-state) regime, the thermal conductivity or *R*-value of a specimen can be obtained. In comparison, the specific heat capacity and thermal diffusivity data can be obtained during the transient regime. Additionally, if heat flux pads are employed in the hot box apparatus, it is possible to monitor the energy that passes through a specimen during thermal cycling. In the case of wall systems, it is of interest to compare the thermal performance of one system against another for the purpose of selecting suitable energy efficient structures. However, the nature of this comparison is not as simple as it would seem. In this study, the thermal performance of several wall systems were compared to evaluate their energy efficiencies relative to one another. Under steady-state conditions, it is valid to simply compare the *R*-values against one another assuming equivalent measurement conditions. However, we are actually interested in the real-world performance of these systems, and in this case, the comparison of *R*-values is not sufficient to fully characterize differences in actual thermal performance. For this reason, a 24-h cyclic thermal loading was employed that mimicked a day/night cycle. By measuring the energy transmitted through the wall, the energy

Manuscript received September 21, 2017; accepted for publication December 4, 2017.
[1]National Brick Research Center–Clemson University, 100 Clemson Research Blvd., Anderson, SC 29625, USA
J. P. S. http://orcid.org/0000-0003-1655-8319
[2]ASTM International Symposium on *Masonry 2018* on June 26, 2018 in San Diego, CA, USA.

required to maintain thermal comfort could be determined. This study aims to characterize the real-world thermal performance of several systems by using this method.

Keywords
R-value, clay masonry, thermal mass

Introduction

One of the simplest ways of characterizing the thermal performance of building materials is by measuring its *R*-value. Testing of small-scale components (12 in. by 12 in. [30.48 cm by 30.48 cm] or smaller) typically is performed by the guarded hot plate method, whereas large-scale testing typically is performed in a hot box apparatus. A traditional hot box operates on the premise of measuring the energy required to maintain a constant temperature difference across a test panel. Thermocouples are most commonly used to measure test panel and air temperatures during steady-state testing. One downside to this type of hot box apparatus is that there is no means of measuring dynamic thermal properties of the wall. In order be able to measure the dynamic response of the wall, both temperatures and heat fluxes need to be measured as a function of time. Keeping this in mind, a hot box, similar to the design specified in ASTM C1363-11, *Standard Test Method for Thermal Performance of Building Materials and Envelope Assemblies by Means of a Hot Box Apparatus* [1], was designed and constructed that incorporates heat flux transducers instead of calculating heat flow [2].

Heat flux transducers (HFTs) for measuring heat flux through a wall section have been used previously to measure the heat flux transmitted through a test panel. These sensors have been used to measure heat flux during a step change in temperature [3] or during natural cyclic conditions open to the outside environment [4–6]. Direct measurement of thermal properties (*R*-value, specific heat capacity, diffusivity) typically is not attempted with HFTs but instead is either calculated from an energy balance or estimated by fitting a solution of Fourier's equation to the measured data and extracting the relevant materials properties from the fitted solution [3].

As stated, *R*-value measurements on their own are not enough to satisfactorily describe real-world performance [7]. In reality, the expected climate, *R*-value, and thermal mass of the wall all contribute to the overall performance [7]. Thermal mass must be taken into account to fully describe thermal performance. The interplay among all of these parameters makes comparisons among wall systems difficult.

Several authors have proposed different methods with which to characterize a wall system's thermal performance. Alterman et al. constructed several mock-up houses constructed with different wall systems, instrumented them, and then left the interior unconditioned [7]. By leaving the interior unconditioned, the temperature inside the structure was directly related to the performance of the building. Looking over multiple day cycles, the percentage of time the interior temperature was within thermal comfort (18°C to 24°C; 65°F to 76°F) was used to evaluate how well each wall system performed. Another method for characterizing the dynamic thermal performance is by a so-called dynamic *R*-value (R_d) [8]. This is a calculated *R*-value over a specified amount of time (24 h, month, or year). When comparing climates that are heat-only or cooling-only, the high mass systems performed similarly to the low mass systems with the same *R*-value. However, once both heating and cooling were required, the high mass system typically exhibited much better thermal performance [8].

One final parameter that has been proposed to characterize dynamic behavior of wall systems is called the T-value [9,10]. Under cyclic loading conditions (with interior temperature uncontrolled), plotting exterior surface temperature versus internal surface temperature results in a hysteresis loop. The angle of the line that connects the two extrema of this loop is the T-value. A perfect insulator would be zero while a perfect conductor would be 45 (slope of 1). These plots are also helpful as they provide a simple graphical comparison of different wall sections.

The motivation for developing this hot box was to be able to measure both steady-state properties via direct measurement using thermocouples as well as HFTs and direct dynamic evaluation of wall systems without relying on theoretical models and assumptions. The combination of these two aspects into one apparatus allows for comparison of different walls in both laboratory and simulated dynamic conditions [2].

Experimental

DESIGN

The design of this hot box was based on guidance from ASTM C1363-11 [1] and EN 1934 [11] but with a few modifications. The frame built to hold the test panels for testing was constructed using 2 in. by 10 in. (3.81 cm by 24.13 cm) studs and 2 in. (5.08 cm) extruded polystyrene (XPS) insulation in alternating layers (five studs and four pieces of insulation). These layers were oriented perpendicular to the direction of heat flow through the sample to minimize heat flow through the support frame. The effective *R*-value of the frame was approximately 8.5 m^2K/W (48 $ft^{2\circ}Fhr/BTU$), which is much higher than any of the test panels being tested in this study. The dimensions of the frame allowed for testing of test panels with a length of 104 in. (2.64 m), height of 72 in. (1.83 m), and thickness up to 15 in. (38.1 cm). Two temperature-controlled chambers rolled up to the two sides of the frame and were secured snugly to prevent air leakage/infiltration from the chambers prior to testing, as shown in **Fig. 1**. This assembly was all contained within a climate-controlled room to prevent variations in ambient temperature, as shown in **Fig. 2**, with the temperature-controlled chambers open.

Air temperature within the two chambers was controlled via computer-controlled Julabo FP51-SL Chillers (Julabo, Allentown, PA) with a working fluid temperature from $-42^\circ C$ to $80^\circ C$ ($-44^\circ F$ to $176^\circ F$). The chillers were each controlled via external temperature probes to maintain the desired air temperature set point

FIG. 1 Cross section of hot box—chambers and sample.

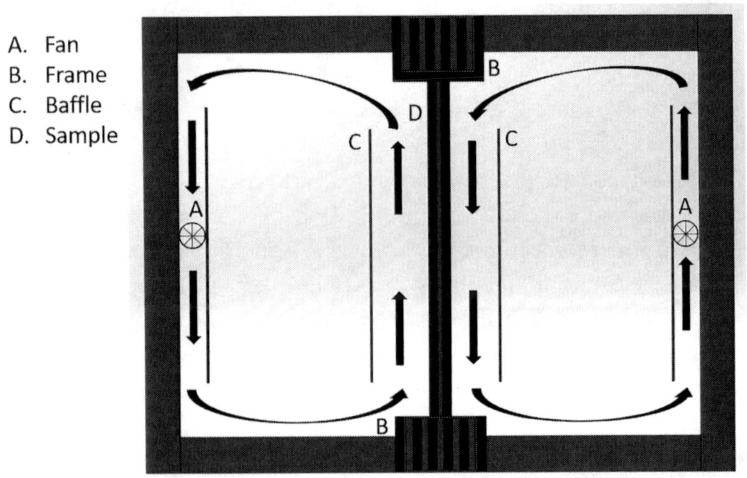

FIG. 2 Hot box showing both chambers, support frame, and climate-controlled room.

rather than using the internal fluid temperature for control; thus, the actual working air temperature range that could be achieved was somewhat less. Heat exchangers were placed in the back of each of the temperature-controlled chambers with a bank of fans providing the airflow as shown in **Fig. 3**. This bank of fans consisted of seven

FIG. 3 Heat exchanger and fan bank.

fans each placed next to one another, which ensured a uniform airflow from the front to the back of the chamber and by extension over the surface of the panel being tested. The fans could be adjusted to modify the airflow rate across the surface of the test panels. Since both chambers were controlled by identical chiller systems, the direction of heat flow through the test panels could be controlled interchangeably. Temperature set points of each chiller were monitored and controlled via computer, which allowed for the use of preprogrammed temperature cycles during dynamic cycling.

Environmental conditions in the hot box climate-controlled room were held constant in order to minimize any extraneous heat flows. The enclosure was kept at 24°C (76°F), and the humidity was controlled below 40 % relative humidity to minimize moisture effects during testing. Typically, one chamber (the metering chamber) was kept at a constant 24°C (76°F), and the temperature in the other (the climactic chamber) was varied. If needed, this configuration could be reversed as stated previously.

TESTING SCHEDULE AND INSTRUMENTATION

Data collection for this hot box was done on a National Instruments PXIe-1078 system (National Instruments, Austin, TX). This system had the capability of simultaneously measuring up to 128 thermocouple inputs, 16 HFTs, 12 analog inputs, as well as controlling both chillers. Not all test panels were instrumented identically but were instrumented as needed in order to fully characterize the test panel based on specific features of the particular wall system. Data were recorded once per second for each input in all tests.

One important assumption on the operation of a hot box is the uniformity of the airflow and temperature over the surface of the test panel. In order to monitor this, 18 thermocouples were spread across the baffle that measured the air temperature near the wall surface. The baffle was used to guide the airflow over the surface of the test panel. The placement of the baffle can be seen in **Fig. 1**. Velocity and humidity measurements were also recorded to ensure both sufficient airflow and that no moisture was condensing during the course of the analysis. This measurement setup was applied to both chambers.

Before testing, or calibration, could begin, a testing schedule had to be established that encompassed both steady-state and dynamic measurements. Four steady-state set points (henceforth referred to as +10, +20, −10, and −20) followed by five 24-h cycles were used for all tests. The steady-state set points were arranged in a step arrangement as shown graphically in **Fig. 4**. For example, in the +10 condition, the air temperature in the climactic chamber was set to 34°C (93°F), or 10°C (18°F) above metering air temperature, and heat flow occurred through the test panel from the climactic side to the metering side. During this testing cycle, the metering chamber was kept at 24°C (76°F) for all parts of the testing cycle. Between each steady-state set point, the test panel was returned to thermal equilibrium with each chamber at 24°C (76°F) as shown in **Fig. 4**.

FIG. 4 Air temperatures during typical testing cycle.

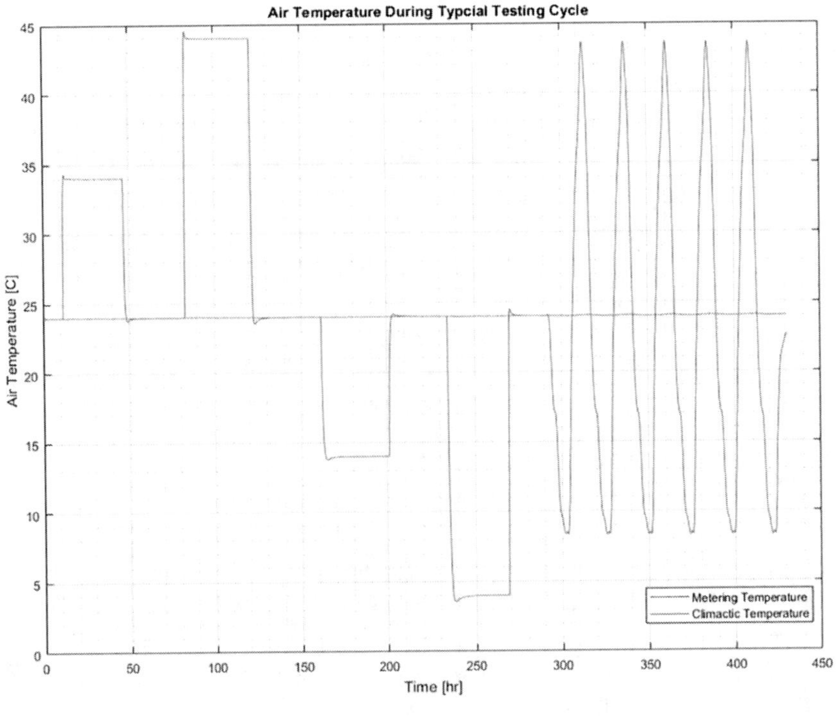

Thermal equilibrium in this setting was defined as when the test panel and both chambers were at the same temperature and there was no net heat flux across the test panel. During each testing set point, the test panel underwent two different thermal regimes: transient and steady-state. The transient regime began as soon as the chiller received the command to change set point and lasted until the test panel reached a state of constant heat flux. When this happened, steady-state conditions were maintained for 25 % of the time that the transient state lasted. This was repeated for each steady-state set point.

After the four steady-state step measurements were made, the dynamic cycle was started and continued for five days. The dynamic cycle that was chosen for this study simulates the day/night cycle in a fairly extreme climate. It is a so-called "sol-air cycle" that was given in a National Bureau of Standards publication in 1972 [12]. The peak temperature for this cycle was approximately 45°C (113°F), and the minimum was approximately 8°C (46°F). Typically, data for the first cycle had to be discarded because the wall was still being influenced by the previous thermal history. Cycles two through five were typically in very good agreement for both temperatures and heat fluxes.

Instrumentation of each wall selected based upon the specific features of each test panel. For example, the calibration materials for this hot box were rather uniform across the entire specimen, so one HFT was placed in the center of the wall on each side and surrounded by nine thermocouples to investigate how uniform the surface temperature was. On the tested wall sections, however, there was an internal wooden (stud) frame with 2 in. by 4 in. (3.81 cm by 8.89 cm) studs every 16 in. (40.64 cm). This complex geometry dictated more complex HFT and thermocouple placement. Four large (12 in. by 12 in.) (30.48 cm by 30.48 cm) HFTs were placed over the insulation, and six small (2 in. by 2 in.) (5.08 cm by 5.08 cm) HFTs were placed over the studs. Twelve thermocouples were affixed to the wall, six over the studs, and six over the insulation (two top, two middle, and two bottom) for each. This was done on the side of the wall with the sheetrock (metering side). On the climactic side (exterior cladding), six medium size (4 in. by 4 in. [10.16 cm by 10.16 cm]) HFTs were affixed to the wall—top, middle, and bottom over the insulation and studs. Thermocouple placement on this side was identical to the metering side.

CALIBRATION

In order to ensure that the results that were measured in our hot box were accurate, and to be able to compare them with results measured by others, extensive calibration of this system had to be performed. Before calibration of *R*-value and *Cp* [2,13] was performed, all thermocouples and HFTs were calibrated individually. HFTs were calibrated by determining their sensitivity coefficients according to ASTM C1130, *Standard Practice for Calibration of Thin Heat Flux Transducers* [14]. Thermocouples were calibrated individually by using an Omega CL25 Digital Calibrator (Omega, Norwalk, CT) whose performance was verified by an accredited metrology laboratory.

Calibration of *R*-value and *Cp* of the system were performed by testing walls made of homogenous material with known properties. For our purposes, we selected three materials: 2 in. (5.08 cm) XPS, 1 in. (2.54 cm) XPS, and plywood. Plywood was chosen because it has a lower *R*-value but a higher *Cp* than the XPS. The R-values of each material were characterized prior to calibration via a Netzsch HFM 436 Lambda Heat Flux Meter (Netzsch, Burlington, MA). Specific heat capacities of XPS were taken from literature sources while the value for plywood was determined by a modified hot box technique that we developed for small sample testing [13]. Nominal values for *R*-value and specific heat capacity are given in **Table 1**. Temperature-dependent *R*-values were needed because this test protocol does not keep constant mean temperature but instead a *constant* air temperature on one side of the test panel. *R*-values as determined by this hot box are compared to the *R*-values as determined by a Netzsch HFM 436 Lambda Heat Flux Meter in **Fig. 5**. The Netzsch

TABLE 1 Calibration materials.

Material	Nominal *R*-Value (@24°C) [m^2K/W] (ft^{2}°Fhr/BTU)	Nominal *Cp* (@24°C) [J/kgK] (BTU/lb°F)
2 in. (5.08 cm) XPS	1.76 (10)	1,280 (0.306)
1 in. (2.54 cm) XPS	0.88 (5)	1,280 (0.306)
Plywood	0.125 (0.71)	1,453 (0.347)

FIG. 5 Measured versus expected *R*-value.

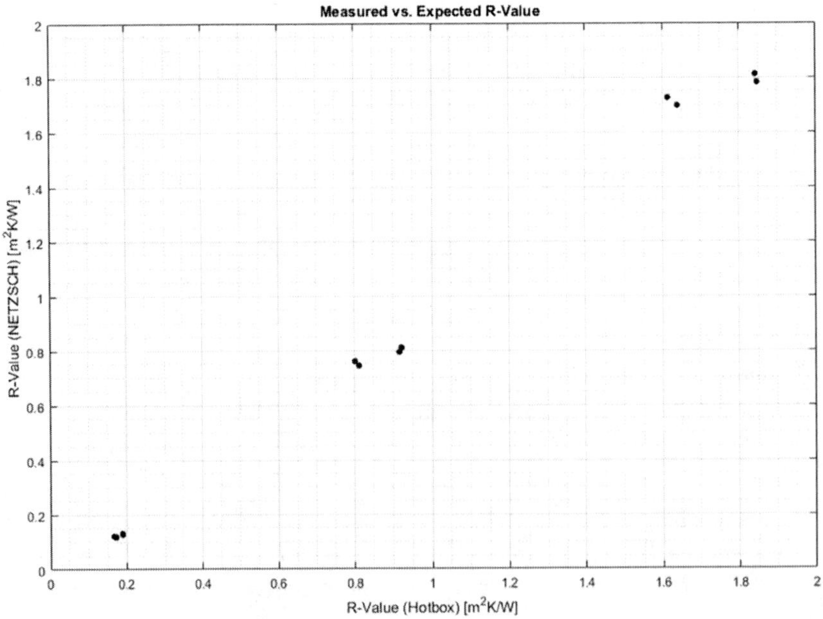

HFM 436 Lambda Heat Flux Meter in **Fig. 5** was calibrated and operated in accordance with ASTM C518, *Standard Test Method for Steady-State Thermal Transmission Properties by Means of The Heat Flow Meter Apparatus* [15].

The *R*-value measurements, shown in **Fig. 5**, for each wall appear to be grouped in pairs. Those pairs correspond to the different directions of heat flow during the previously described set of temperature measurements (**Fig. 4**). This difference in heat flow direction necessitated two calibration equations, one for each direction of heat flow. Also, since the mean temperature was not held constant and varied from 14°C to 34°C (57°F to 93°F), this also had to be incorporated in the calibration equation.

After analysis of a few different models, the following calibration equation (Eq 1) was chosen due to the small number of coefficients and simple form:

$$R_N = (aR_{HB}^b)(1 + cT) \quad (1)$$

where:

R_N = *R*-value as measured by Netzsch,
R_{HB} = *R*-value as measured by hot box, and
T = average temperature of the wall in °C.

Performing nonlinear regression on the two data sets separately yielded Eq 2 and Eq 3:

$$R_N = \left(1.186\, R_{HB}^{1.172}\right)(1 - 0.006T) \quad (2)$$

$$R_N = \left(0.938\, R_{HB}^{1.157}\right)(1 - 0.004T) \quad (3)$$

The corrected *R*-value data are plotted against the Netzsch HFM 436 Lambda Heat Flux Meter *R*-values in **Fig. 6**, and a comparison is given in **Table 2**.

After performing calibration of *R*-value, calibration of specific heat capacity was attempted. The heat capacity for this hot box was calibrated in the same manner as that of the modified hot box to measure specific heat capacity [13]. The specific heat capacity of the test panel is given by the area between the two heat flux curves—essentially the difference between the energy that went into the wall versus what came out. This technique of measuring heat capacity was described previously [2]. The measured amount of energy storage versus the expected storage based upon specific heat capacity, thickness, and bulk density of the test panels is given in **Fig. 7**.

In order to calibrate these measurements, the test panels were again separated into pairs as before based upon the direction of heat flow. In analyzing the results of these measurements, temperature again had to be taken into account because the mean temperature was not constant. The following model (Eq 4) was selected due to good agreement with the observed data as well as the small number of coefficients:

$$E_T = a + bT + cE_M \quad (4)$$

where:

E_T = true energy storage of the wall,
T = average temperature of the wall in °C, and
E_M = measured energy storage.

FIG. 6 Corrected versus expected *R*-value.

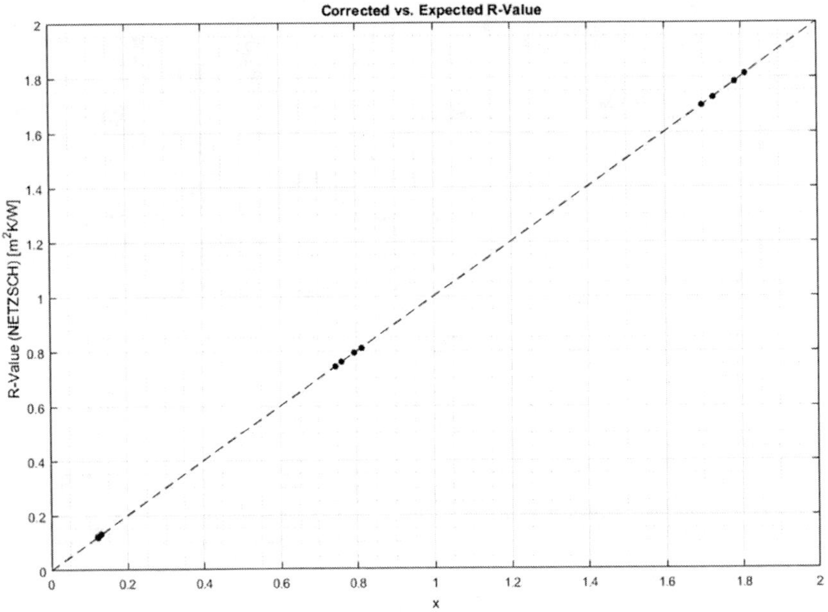

TABLE 2 Comparison of corrected and Netzsch HFM 436 lambda *R*-values.

Material	Temp [°C] (°F)	*R*-Netzsch [m²K/W] (ft²°Fhr/BTU)	*R*-Corrected [m²K/W] (ft²°Fhr/BTU)	% Difference
2 in. XPS	13.73 (56.71)	1.812 (10.29)	1.811 (10.28)	−0.05 %
	18.67 (65.61)	1.783 (10.12)	1.783 (10.12)	0.02 %
	28.36 (83.05)	1.725 (9.8)	1.726 (9.8)	0.02 %
	33.2 (91.76)	1.697 (9.64)	1.696 (9.63)	−0.02 %
1 in. XPS	13.77 (56.79)	0.81 (4.6)	0.813 (4.62)	0.31 %
	18.62 (65.52)	0.794 (4.51)	0.793 (4.5)	−0.10 %
	28.37 (83.07)	0.76 (4.32)	0.76 (4.32)	−0.09 %
	33.24 (91.83)	0.744 (4.22)	0.744 (4.22)	0.08 %
Plywood	14.43 (57.97)	0.132 (0.75)	0.13 (0.74)	−1.82 %
	18.97 (66.15)	0.129 (0.73)	0.128 (0.73)	−0.75 %
	28.07 (82.53)	0.122 (0.69)	0.12 (0.68)	−1.63 %
	32.6 (90.68)	0.119 (0.68)	0.121 (0.69)	1.89 %
RMSE				0.92 %

Note: RMSE = root-mean-squared-error.

FIG. 7 Measured versus expected energy storage.

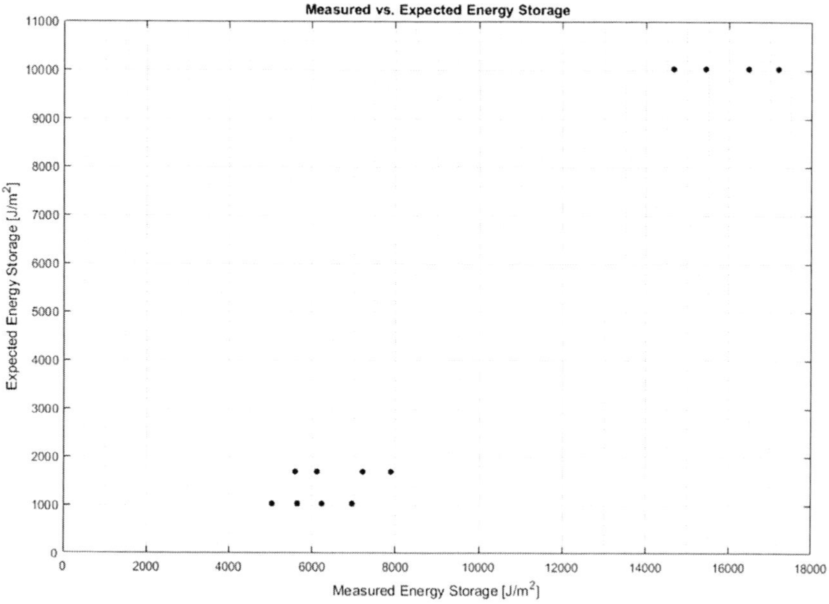

Performing linear regression yielded Eq 5 and Eq 6:

$$E_T = -32.20 - 120.9T + 0.912E_M \tag{5}$$

$$E_T = -2825 - 132.1T + 0.894E_M \tag{6}$$

Finally, the corrected energy storage was compared against the expected energy storage, and the results of this comparison are shown graphically in **Fig. 8**. After calculating the specific heat capacity from the corrected energy storage, these values were compared against the specific heat capacity data that were previously reported in **Table 1**. This comparison of measured specific heat capacity to expected heat capacity is reported in **Table 3**. Good agreement between the expected and measured values was obtained using this calibration method.

COMPARISON OF WALL SYSTEMS

For the comparison of different wall systems, four different test panels sections were constructed as they would be expected to be built in practice. The wall types primarily differed in the external cladding, which was as follows: fiber cement board, vinyl siding, brick veneer (open and closed weeps), and stud wall (no cladding). A description of each wall is given in **Table 4**. The brick used to construct the veneer test specimens was a modular-sized brick that met ASTM C216, *Standard Specification for Facing Brick (Solid Masonry Units Made from Clay or Shale)* [16]. The stud wall with no cladding was tested and used as a baseline comparison to

FIG. 8 Corrected versus expected energy storage.

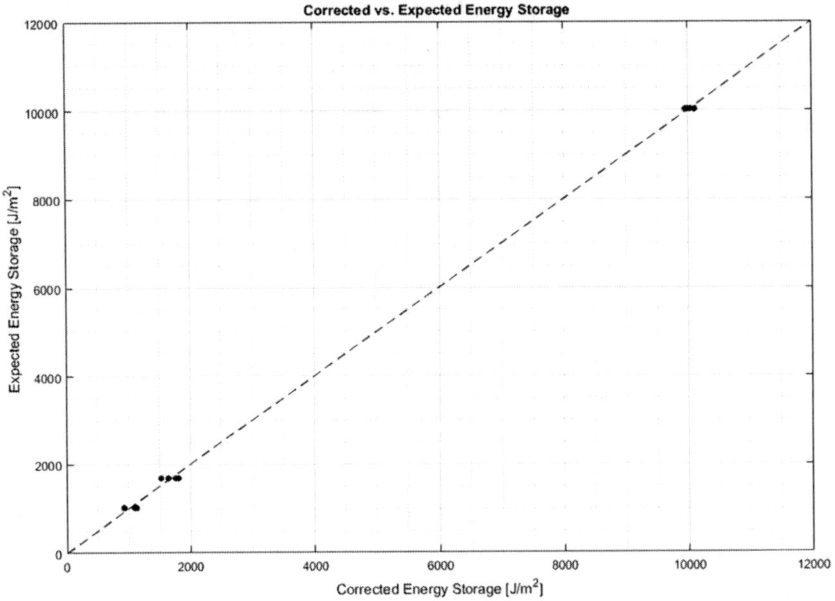

TABLE 3 Comparison of corrected *Cp* and expected *Cp* values.

Material	Temp [°C] (°F)	*Cp* Expected [J/kgK] (BTU/lb°F)	*Cp* Measured [J/kgK] (BTU/lb°F)	% Difference
2 in. XPS	13.73 (56.71)	1,280 (0.306)	1,374 (0.328)	7.35 %
	18.67 (65.61)	1,280 (0.306)	1,339 (0.32)	4.57 %
	28.36 (83.05)	1,280 (0.306)	1,250 (0.299)	−2.36 %
	33.2 (91.76)	1,280 (0.306)	1,162 (0.278)	−9.23 %
1 in. XPS	13.77 (56.79)	1,280 (0.306)	1,161 (0.277)	−9.31 %
	18.62 (65.52)	1,280 (0.306)	1,172 (0.28)	−8.46 %
	28.37 (83.07)	1,280 (0.306)	1,414 (0.338)	10.46 %
	33.24 (91.83)	1,280 (0.306)	1,380 (0.33)	7.83 %
Plywood	14.43 (57.97)	1,453 (0.347)	1,448 (0.346)	−0.32 %
	18.97 (66.15)	1,453 (0.347)	1,455 (0.348)	0.13 %
	28.07 (82.53)	1,453 (0.347)	1,443 (0.345)	−0.72 %
	32.6 (90.68)	1,453 (0.347)	1,465 (0.35)	0.80 %
RMSE				6.43 %

determine the relative effect of each type of cladding. Each wall was instrumented as mentioned earlier in an effort to characterize both the portion of the wall with insulation and that with the studs. These wall systems were chosen because they

TABLE 4 Summary of construction of test panels.

Test Panel	Cladding	Air Space	House Wrap	Stud	Insulation	Interior
Stud Wall	None	None	7/16 in. (1.11 cm) OSB	2 by 4 (3.81 cm by 8.89 cm)	*R*-13 (2.29 m^2K/W)	½ in. (1.27 cm) gypsum
Fiber Cement	fiber cement board	air behind overlapped panels	7/16 in. (1.11 cm) OSB + Tyvek	2 by 4 (3.81 cm by 8.89 cm)	*R*-13 (2.29 m^2K/W)	½ in. (1.27 cm) gypsum
Vinyl Siding	vinyl siding	air under pocket formed by vinyl	7/16 in. (1.11 cm) OSB + Tyvek	2 by 4 (3.81 cm by 8.89 cm)	*R*-13 (2.29 m^2K/W)	1/2 in. (1.27 cm) gypsum
Brick Wall	3.5 in. (8.89 cm) modular cored brick + mortar	1 in. air cavity	7/16 in. (1.11 cm) OSB + Tyvek	2 by 4 (3.81 cm by 8.89 cm)	*R*-13 (2.29 m^2K/W)	1/2 in. (1.27 cm) gypsum

Note: OSB = oriented strand board.

represent some of the most common residential construction systems in the United States. The measured data for each wall are given in Appendix A.

One interesting thing to note from the results of this comparison was that the stud and insulation parts of the wall behave very differently. This was somewhat expected because the studs act as a thermal bridge and bypass the insulation. Interestingly, the surface temperature of the test panels was not uniform even though a uniform air temperature was applied. In all cases for each test panel, there was a smaller temperature gradient across the studs than was found across the insulation due to higher heat flow through the studs. This higher heat flux was a direct result of the studs acting as a thermal bridge.

As expected, the stud wall had the lowest *R*-value and highest heat flux through both the insulation and studs. One caveat to the results from the vinyl siding test panel was that since the HFTs must be affixed to a flat surface, only the wall over the air pocket created by the overlap of the vinyl sheets as shown in **Fig. 9** could be tested. This air pocket resulted in poor thermal contact between the HFT and the rest of the wall system. For this reason, the actual wall system is likely to perform slightly worse than presented here.

In calculating the average *R*-value for the test panel, the *R*-values over the stud and over the insulation were first calculated, then averaged. The average *R*-value for each test panel is given in **Table 5**. All *R*-values that are reported in this study are R_{SI} values, which have the units of $m^2 \cdot K/W$. The English equivalent of *R*-value has the units $ft^2 \cdot {}^\circ F \cdot hr/BTU$. The relationship between R_{SI} and R is given (approximately) by Eq 7.

$$R = R_{SI} \times 5.678 \qquad (7)$$

FIG. 9 Air pocket created by overlap of vinyl sheets.

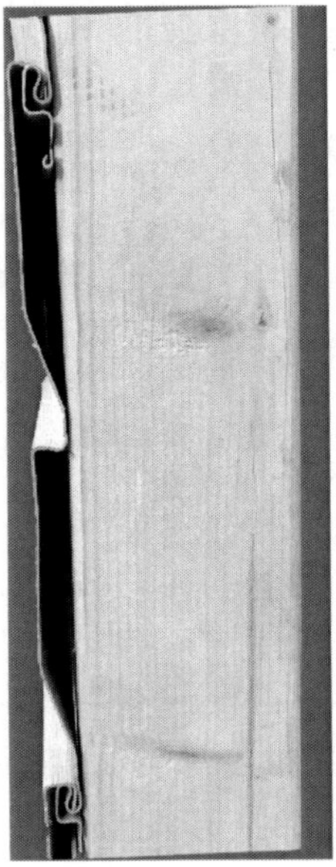

TABLE 5 Comparison of test panel average R_{SI}-values [m^{2}K/W] (BTU/lb°F).

Condition	Stud	Fiber Cement	Vinyl	Brick Veneer–Open	Brick Veneer–Closed
+10	1.88 (10.7)	1.98 (11.2)	2.25 (12.8)	1.98 (11.2)	2.18 (12.4)
+20	1.82 (10.3)	1.87 (10.6)	2.12 (12)	2.06 (11.7)	2.03 (11.5)
−10	2.02 (11.5)	2.13 (12.1)	2.31 (13.1)	2.31 (13.1)	2.21 (12.5)
−20	2.08 (11.8)	2.15 (12.2)	2.37 (13.5)	2.26 (12.8)	2.21 (12.5)
Average R_{SI}	1.95 (11.1)	2.03 (11.5)	2.26 (12.8)	2.15 (12.2)	2.16 (12.3)

The heat flux used to calculate the R_{SI} value comes from an average of the climactic and metering heat fluxes for each respective area. One way to account for differing claddings that are not well suited to instrumentation by HFTs, such as the vinyl, was to use only the heat flux from the metering wall surface in the *R*-value calculations. These data are presented in **Table 6** and are termed metering R_{SI}-value.

On most walls, we see a difference on the climactic side over the studs versus over the insulation. This effect was significantly reduced for the brick veneer wall. This was likely due to the air cavity and brick layer homogenizing the heat flux somewhat. Looking at average *R*-values, the steady-state thermal performance of the test panel with vinyl was slightly better than the brick veneer test panels, but

TABLE 6 Comparison of test panel metering R_{SI}-values [m^2K/W] (BTU/lb°F).

Condition	Stud	Fiber Cement	Vinyl	Brick Veneer-Open	Brick Veneer-Closed
+10	2.04 (11.6)	2.05 (11.6)	2.16 (12.3)	2.31 (13.1)	2.24 (12.7)
+20	1.87 (10.6)	1.96 (11.1)	2.03 (11.5)	2.17 (12.3)	2.16 (12.3)
−10	2.07 (11.8)	2.13 (12.1)	2.18 (12.4)	2.26 (12.8)	2.3 (13.1)
−20	2.07 (11.8)	2.16 (12.3)	2.21 (12.5)	2.41 (13.7)	2.34 (13.3)
Average R_{SI}	2.01 (11.4)	2.08 (11.8)	2.14 (12.2)	2.29 (13)	2.26 (12.8)

when looking at the metering *R*-value, the brick veneer panel actually performed better with respect to steady-state *R*-value measurements. This discrepancy is likely due to what was described earlier, where the HFTs could only be placed over the air pockets on the vinyl wall. This appears to show that the brick veneer test panel performed better than the vinyl test panel. Also note that there does not appear to be any significant difference between open and closed weep configurations on the brick wall in either average or metering R_{SI}-values.

In all cases, the cladding improved the thermal performance of the basic stud wall that was common to all of these test panels. The measured *R*-values of the brick veneer test panels were higher than the fiber cement. Movement of air through the weeps (open versus closed conditions) did not appear to affect the average R_{SI}-values of the brick veneer test panel. Further studies of air movement in the air space are ongoing.

By evaluating the transient regime data that were collected during the steady-state step measurements shown in **Fig. 4**, thermal storage data can be extracted. The heat flux difference curve (difference between heat flux into and out of the test panel) is shown in **Fig. 10** for each test panel. Before the step began, there was no net flux, and once it had reached a constant heat flux transmittance (steady-state), there was no net heat flux again.

Although each of the walls had similar specific heat capacities (*Cp*) (1,000–1,400 J/kgK; 0.239–0.334 BTU/lb°F) as reported in Appendix A, the actual amount of heat storage (heat capacity) was very different and is given in **Table 7**. This was due to differences in bulk density for each of the test panels. Converting specific heat capacity (J/kgK) into heat capacity (J/K) highlights the differences among the wall systems.

The test panels that had a larger thermal mass (area under the curve in **Fig. 10**) also had larger heat capacities. From these plots, we can also extract the amount of time required to reach equilibrium for each system, which is a further indication of the effect of thermal mass. This time comparison is reported in **Table 8**.

From the data reported in **Table 8**, the argument can be made that the brick wall will have a much more substantial lag time under cyclic conditions than the other test panels. In this study, the interior air temperature was kept at a constant 24°C throughout the dynamic cycles. Instead of computing lag times, the total

FIG. 10 Heat flux difference curve during a single temperature step.

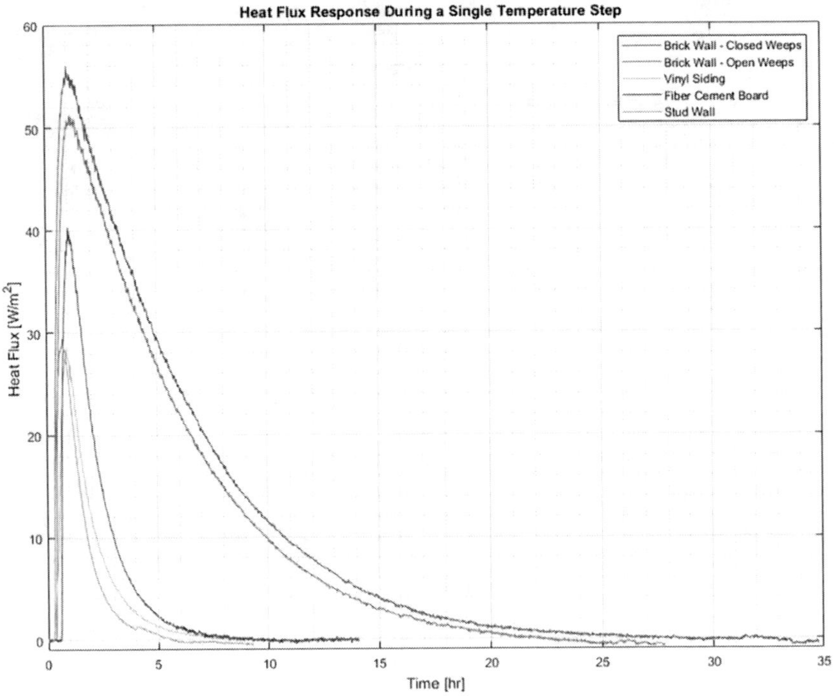

TABLE 7 Heat capacity of different walls [J/K] (BTU/°F).

Condition	Stud	Fiber Cement	Vinyl	Brick Veneer–Open	Brick Veneer–Closed
+10	109 (0.057)	216 (0.114)	138 (0.073)	974 (0.513)	1,128 (0.594)
+20	126 (0.066)	214 (0.113)	139 (0.073)	1,116 (0.588)	1,111 (0.585)
−10	112 (0.059)	211 (0.111)	122 (0.064)	1,110 (0.585)	1,085 (0.571)
−20	117 (0.062)	207 (0.109)	125 (0.066)	990 (0.521)	1,090 (0.574)
Average	116 (0.061)	212 (0.112)	131 (0.069)	1,048 (0.552)	1,104 (0.581)

TABLE 8 Time to reach equilibrium for different walls (hours).

Condition	Stud	Fiber Cement	Vinyl	Brick–Open	Brick–Closed
+10	5.9	7.3	8.3	22.6	22.6
+20	6.4	8.5	8.1	25.5	24.2
−10	6.3	7.9	7.8	22.5	22.1
−20	6.7	8.1	8.1	24.7	25.4
Average	6.4	7.9	8.1	23.8	23.6

amount of energy transmitted through the test panel during a given cycle was measured to simulate the load on an HVAC system that would be required to maintain a constant internal temperature. Heat fluxes during the dynamic cycle for the stud wall are shown in **Fig. 11**. Note from the figure that the amplitude of the heat fluxes over the studs was larger for both the climactic and metering surfaces. Also, there was some additional lag on the metering wall heat flux caused by the larger heat storage in that portion of the test panel. Note: A positive heat flux on the metering side and a negative heat flux on the climactic side indicate the same direction of heat flow.

It should be noted that all of the test panels in this study were subjected to the same dynamic cycling conditions as described previously. Any comparison of dynamic thermal performance is relative to the cycle that has been tested, and cycles with large temperature swings tend to highlight the beneficial effects of thermal mass on reducing heat flow.

To allow for a direct comparison of the wall systems, the heat fluxes over the studs and insulation on the metering side (interior surface) were simply averaged and integrated from one zero heat flux point to the next zero heat flux point as described previously. A comparison plot of the metering wall heat flux is shown in **Fig. 12**. The metering wall heat flux represents the amount of energy that makes it

FIG. 11 Heat fluxes for stud wall during dynamic cycle.

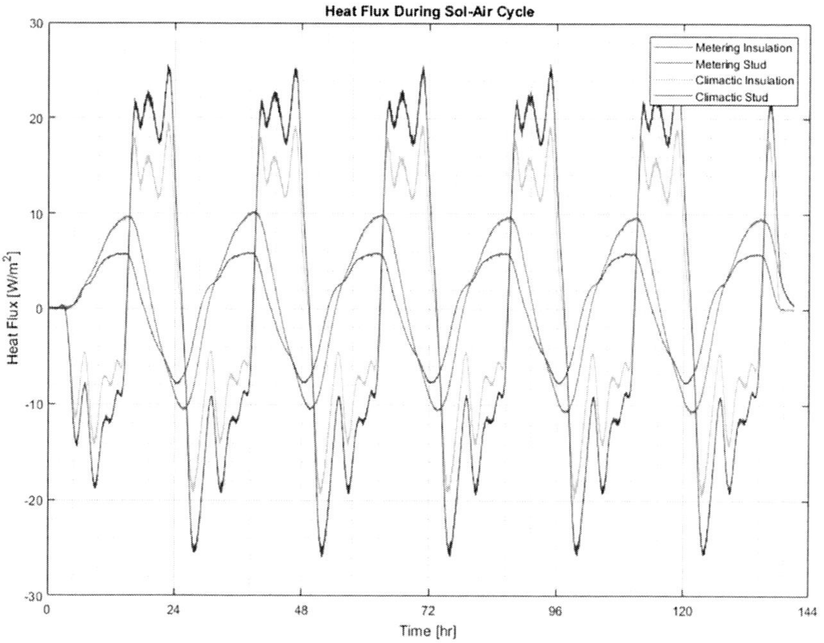

FIG. 12 Metering wall heat flux during dynamic cycling.

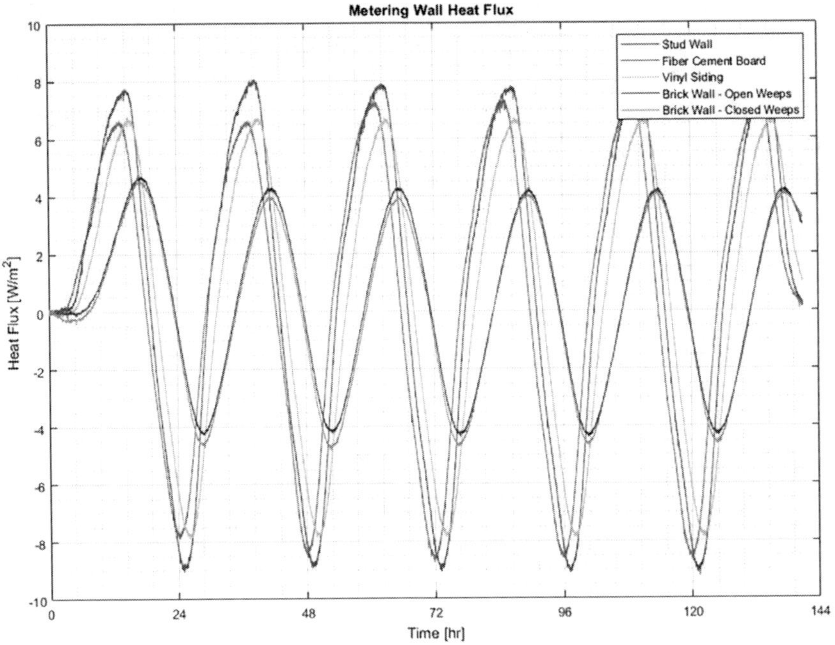

through the wall and into the metering chamber. This would simulate the load on an HVAC system for a given wall system. As can be seen in **Fig. 12**, the vinyl and fiber cement board test panels performed slightly better than the stud wall, but the brick wall did significantly better due to the larger thermal mass. **Table 9** gives the average amount of energy that passed through the wall for each cycle. This was done by integrating the curves in **Fig. 12** for each cycle then averaging the results for several cycles. A positive heat flow direction, as indicated in **Table 9**, means that there was heat flow into the wall from the metering chamber and that the climactic side of the test panel was much colder. A negative heat flow direction means that there was heat flow from the wall into the metering chamber and that the climactic side of the test panel was hotter.

TABLE 9 Total energy transmitted (average area under curve) [J/m^2].

Direction	Stud Wall	Fiber Cement	Vinyl	Brick-Open	Brick-Closed
Positive	2.38E+05	2.08E+05	1.90E+05	1.16E+05	1.05E+05
Negative	−2.07E+05	−1.93E+05	−1.87E+05	−1.07E+05	−1.19E+05

Using this simulated cycle, there was a much lower amount of energy transmitted through the brick veneer test panel as compared to the other three wall system test panels that were evaluated in this study. This reduction in heat flow was directly related to the thermal mass acting as a damper to the amount of heat transmitted through the wall. Again, very little difference between the open and closed weep configuration of the brick veneer test panel was noted.

In order to quantitatively compare the dynamic thermal performance of each test panel subjected to this particular dynamic cycling condition, a percent increase in *R*-value relative to the stud wall was calculated and is summarized in **Table 10**. In order to compare dynamic data, a percent reduction in total energy transmitted was calculated relative to the stud wall and is reported in **Table 11**. When comparing the relative performance of each test panel relative to the stud wall, it is apparent that dynamic cycling yielded better than expected thermal performance compared to the steady-state measurements. This was a direct result of the thermal mass of the test panels. These results are valid for the comparison against one another under the same cyclic conditions. These results do not necessarily reflect how each test panel will perform under different cyclic conditions. The steady-state and dynamic measurements in this study likely represent a range of expected performance for each of these wall systems. In other words, the steady-state *R*-value measurements provide a measurement of the minimum performance, while the dynamic measurements highlight the potential for additional reductions in energy consumptions under favorable atmospheric conditions.

TABLE 10 Thermal performance relative to stud wall (percent increase using metering R_{SI}-value).

Condition	Fiber Cement	Vinyl	Brick Veneer-Open	Brick Veneer-Closed
+10	0.5 %	5.5 %	13.0 %	9.5 %
+20	5.1 %	8.7 %	16.4 %	15.5 %
−10	2.5 %	5.1 %	8.8 %	11.0 %
−20	4.2 %	6.7 %	16.5 %	13.0 %
Average	3.1 %	6.5 %	13.7 %	12.3 %

TABLE 11 Thermal performance relative to stud wall (percent decrease in transmitted energy).

Direction	Fiber Cement	Vinyl	Brick-Open	Brick-Closed
Positive	12.6 %	20.3 %	51.1 %	55.8 %
Negative	6.7 %	9.5 %	48.4 %	42.6 %
Average	9.6 %	14.9 %	49.8 %	49.2 %

Conclusions

Although R_{SI}-value determination is a simple method to characterize the thermal performance of a wall system, it does not adequately characterize the real-world, dynamic behavior of a wall system. The assumption that a wall system behaves identically under dynamic loading as it does under steady-state conditions has been shown to be false. In order to be able to predict how a given system will behave under dynamic conditions, the thermal mass of the system must be taken into account. Unfortunately, this type of analysis is much more complicated. For example, a simple R-value comparison would indicate that the vinyl cladding wall system performed better than the brick veneer wall system. However, when comparing them under this dynamic cycle, the brick veneer wall system transmitted only half the amount of energy relative to the vinyl cladding wall system. It should always be noted that dynamic thermal performance data are dependent on the actual cycle used for the evaluation. Thermal mass has a much more profound impact on thermal performance than is currently realized but is dependent on local climatic conditions.

References

[1] ASTM C1363-11, *Standard Test Method for Thermal Performance of Building Materials and Envelope Assemblies by Means of a Hot Box Apparatus*, ASTM International, West Conshohocken, PA, 2011, www.astm.org

[2] Sanders, J., Carrier, J., Clark, C., Jr., and Huygen, N., "Dynamic Thermal Performance Measurements of Clay Brick Masonry," presented at the *Thirteenth Canadian Masonry Symposium*, Halifax, Nova Scotia, June 4–7, 2017.

[3] Robinson, A. J., Lesage, F. J., Reilly, A., McGranaghan, G., Byrne, G., O'Hegarty, R., and Kinnane, O., "A New Transient Method for Determining Thermal Properties of Wall Sections," *Energ. Buildings*, Vol. 142, 2017, pp. 139–146, http://dx.doi.org/10.1016/j.enbuild.2017.02.029

[4] Aste, N., Angelotti, A., and Buzzetti, M., "The Influence of the External Walls Thermal Inertia on the Energy Performance of Well Insulated Buildings," *Energ. Buildings*, Vol. 41, No. 11, 2009, pp. 1181–1187, http://dx.doi.org/10.1016/j.enbuild.2009.06.005

[5] Biddulph, P., Gori, V., Elwell, C. A., Scott, C., Rye, C., Lowe, R., and Oreszczyn, T., "Inferring the Thermal Resistance and Effective Thermal Mass of a Wall Using Frequent Temperature and Heat Flux Measurements," *Energ. Buildings*, Vol. 78, 2014, pp. 10–16, http://dx.doi.org/10.1016/j.enbuild.2014.04.004

[6] Pernigotto, G., Prada, A., Patuzzi, F., Baratieri, M., and Gasparella, A., "Characterization of the Dynamic Thermal Properties of the Opaque Elements Through Experimental and Numerical Tests," *Enrgy. Proced.*, Vol. 78, 2015, pp. 3234–3239, http://dx.doi.org/10.1016/j.egypro.2015.11.786

[7] Alterman, D., Page, A. W., Moghtaderi, B., Zhang, C., and Moffiet, T., "The Influence of Thermal Resistance and Thermal Mass on the Seasonal Performance of Walling Systems in Australia," *J. Green Build.*, Vol. 10, No. 4, 2015, pp. 107–126, http://dx.doi.org/10.3992/jgb.10.4.107

[8] Al-Sanea, S. A., Zedan, M. F., and Al-Hussain, S. N., "Effect of Thermal Mass on Performance of Insulated Building Walls and the Concept of Energy Savings Potential," *Appl. Energ.*, Vol. 89, No. 1, 2012, pp. 430–442, http://dx.doi.org/10.1016/j.apenergy.2011.08.009

[9] Alterman, D., Moffiet, T., Hands, S., Page, A., Luo, C., and Moghtaderi, B., "A Concept for a Potential Metric to Characterise the Dynamic Thermal Performance of Walls," *Energ. Buildings*, Vol. 54, 2012, pp. 52–60, http://dx.doi.org/10.1016/j.enbuild.2012.08.006

[10] Alterman, D., Page, A. W., Zhang, C., and Moghtaderi, B., "The Dynamic Thermal Performance of Masonry Walling Systems," presented at the *Thirteenth Canadian Masonry Symposium*, Halifax, Nova Scotia, June 4–7, 2017.

[11] EN 1934:1998, *Thermal Performance of Buildings—Determination of Thermal Resistance by Hot Box Method Using Heat Flow Meter—Masonry*, European Committee for Standardization, Brussels, Belgium, 1998.

[12] Peavy, B. A., Powell, F. J., and Burch, D. M., *Dynamic Thermal Performance of an Experimental Masonry Building*, Building Science Series 45, US Department of Commerce, National Bureau of Standards, Washington, DC, 1973.

[13] Huygen, N. and Sanders, J., "Dynamic Thermal Performance Measurements of Residential Wall Systems," *Masonry 2018, ASTM STP1612*, N. V. Krogstad and W. M. McGinley, Eds., ASTM International, West Conshohocken, PA, 2018, pp. 83–106, http://dx.doi.org/10.1520/STP161220170173

[14] ASTM C1130-17, *Standard Practice for Calibrating Thin Heat Flux Transducers*, ASTM International, West Conshohocken, PA, 2017, www.astm.org

[15] ASTM C518-17, *Standard Test Method for Steady-State Thermal Transmission Properties by Means of The Heat Flow Meter Apparatus*, ASTM International, West Conshohocken, PA, 2017, www.astm.org

[16] ASTM C216-17, *Standard Specification for Facing Brick (Solid Masonry Units Made from Clay or Shale)*, ASTM International, West Conshohocken, PA, 2017, www.astm.org

Appendix A—Thermal Performance Measurements of Residential Walls

TABLE A1 Stud wall.

		Steady-State				
Property	**Unit**	**+10**	**+20**	**−10**	**−20**	**Average**
TMI	°C	24.05	24.43	23.32	23.00	
TMS	°C	24.23	24.80	23.15	22.63	
TCI	°C	32.94	42.27	13.80	4.26	
TCS	°C	32.81	42.02	13.97	4.58	
QMI	W/m^2	4.0	8.3	3.6	7.3	
QMS	W/m^2	5.7	12.4	5.8	11.7	
QCI	W/m^2	4.7	9.0	4.0	7.6	
QCS	W/m^2	6.4	12.2	5.5	10.7	
R-Avg	m^{2}K/W	1.88	1.82	2.02	2.08	1.95
R-Met	m^{2}K/W	2.04	1.87	2.07	2.07	2.01
Cp	J/kgK	1,177	1,355	1,203	1,258	1,248

Note: TMI = metering surface temperature over insulation; TMS = metering surface temperature over stud; TCI = climactic surface temperature over insulation; TCS = climactic surface temperature over stud; QMI = heat flux on metering surface over insulation; QMS = heat flux on metering surface over stud; QCI = heat flux on climactic surface over insulation; QCS = heat flux on climactic surface over stud; R-avg = average R-value of the wall (average of R-value over stud and over insulation); R-met = R-value calculated using only metering side heat fluxes; Cp = average specific heat capacity (averaged heat flux signals before integrating).

TABLE A2 Fiber cement board.

		Steady-State				
Property	**Unit**	**+10**	**+20**	**−10**	**−20**	**Average**
TMI	°C	24.18	24.63	23.36	22.98	
TMS	°C	24.28	24.83	23.25	22.77	
TCI	°C	32.73	42.03	13.98	4.55	
TCS	°C	32.65	41.90	14.07	4.69	
QMI	W/m^2	3.7	7.7	3.5	6.9	
QMS	W/m^2	5.8	11.8	5.8	11.3	
QCI	W/m^2	4.4	9.2	3.9	7.8	
QCS	W/m^2	5.3	11.1	4.7	9.4	
R-Avg	m^{2}K/W	1.98	1.87	2.13	2.15	2.03
R-Met	m^{2}K/W	2.05	1.96	2.13	2.16	2.08
Cp	J/kgK	1,415	1,402	1,385	1,362	1,391

TABLE A3 Vinyl siding

		Steady-State				
Property	**Unit**	**+10**	**+20**	**−10**	**−20**	**Average**
TMI	°C	24.03	24.38	23.34	23.03	
TMS	°C	24.22	24.76	23.18	22.69	
TCI	°C	32.83	42.18	13.91	4.43	
TCS	°C	32.76	42.05	13.99	4.60	
QMI	W/m^2	3.7	7.6	3.4	6.8	
QMS	W/m^2	5.7	11.8	5.7	11.2	
QCI	W/m^2	3.7	7.7	3.4	6.6	
QCS	W/m^2	4.6	9.4	4.3	8.3	
R-Avg	m^{2}K/W	2.25	2.12	2.31	2.37	2.26
R-Met	m^{2}K/W	2.16	2.03	2.18	2.21	2.14
Cp	J/kgK	1,388	1,405	1,227	1,260	1,320

TABLE A4 Brick veneer and stud wall (open weeps).

		Steady-State				
Property	**Unit**	**+10**	**+20**	**−10**	**−20**	**Average**
TMI	°C	24.08	24.41	23.38	23.05	
TMS	°C	24.20	24.66	23.27	22.84	
TCI	°C	32.74	42.19	13.92	4.41	
TCS	°C	32.64	41.96	14.07	4.66	
QMI	W/m^2	3.4	7.0	3.3	6.3	
QMS	W/m^2	5.5	11.5	5.6	10.6	
QCI	W/m^2	4.9	8.8	3.7	8.1	
QCS	W/m^2	5.5	10.0	4.1	9.1	
R-Avg	m^{2}K/W	1.98	2.06	2.31	2.26	2.15
R-Met	m^{2}K/W	2.31	2.17	2.26	2.41	2.29
Cp	J/kgK	1,055	1,209	1,202	1,072	1,134

TABLE A5 Brick veneer and stud wall (closed weeps).

		Steady-State				
Property	**Unit**	**+10**	**+20**	**−10**	**−20**	**Average**
TMI	°C	24.09	24.49	23.38	23.04	
TMS	°C	24.21	24.73	23.28	22.84	
TCI	°C	32.73	42.13	13.97	4.56	
TCS	°C	32.66	41.96	14.07	4.73	
QMI	W/m^2	3.4	7.0	3.3	6.4	
QMS	W/m^2	5.6	11.5	5.4	10.8	
QCI	W/m^2	4.1	9.0	4.1	8.4	
QCS	W/m^2	4.6	9.9	4.4	8.6	
R-Avg	m^2K/W	2.18	2.03	2.21	2.21	2.16
R-Met	m^2K/W	2.24	2.16	2.30	2.34	2.26
Cp	J/kgK	1,222	1,203	1,175	1,180	1,195

STP 1612, 2018 / available online at www.astm.org / doi: 10.1520/STP161220170171

Ian R. Chin[1]

Brick Masonry in Chicago: Crisis in Aesthetics?

Citation

Chin, I. R., "Brick Masonry in Chicago: Crisis in Aesthetics?," *Masonry 2018*, *ASTM STP1612*, N. V. Krogstad and W. M. McGinley, Eds., ASTM International, West Conshohocken, PA, 2018, pp. 107–122, http://dx.doi.org/10.1520/STP161220170171[2]

ABSTRACT

The Great Chicago Fire lasted for almost three days and affected a territory about 4 miles (6.5 km) in length by an average of two-thirds of a mile (1 km) in width (burnt district). The fire destroyed about 18,000 mostly wood-framed buildings. Soon after the fire, rebuilding began in earnest, and about 100 years after the fire, the world's tallest building was constructed in the burnt district. Many of the buildings that were constructed in the burnt district during the period of 1871 to 1930, or approximately 60 years after the fire had brick masonry exterior walls. Several of these buildings utilized innovative, exciting, and aesthetically pleasing applications of brick masonry wall systems. During the period of 1930 to 2017, or approximately 60 to 150 years after the fire, no significant building with innovative, exciting, and aesthetically pleasing applications of brick masonry wall systems was constructed in the burnt district. This condition led the author to consider what eventually became the title of this paper: *Brick Masonry in Chicago: Crisis in Aesthetics?* Based upon the evaluation of buildings that were constructed with brick masonry exterior walls in the burnt district from 1871 to 2017, or approximately 150 years subsequent to the fire, it is the author's opinion that there is an aesthetic crisis in these buildings and that improvements in the aesthetics should be considered. Recommendations to improve the aesthetics of buildings with brick masonry in exterior wall systems are presented in this paper for review and consideration by interested parties. The results of this evaluation are pertinent to buildings constructed in the burnt

Manuscript received September 19, 2017; accepted for publication December 12, 2017.

[1]Wiss, Janney, Elstner Associates, Inc., 10 South LaSalle St., Suite 2600, Chicago, IL 60603, USA

http://orcid.org/0000-0003-3633-1576

[2]ASTM International Symposium on *Masonry 2018* on June 26, 2018 in San Diego, CA, USA.

district after the fire and may not apply to buildings constructed with brick masonry exterior walls elsewhere during this same time period.

Keywords
Great Chicago Fire, brick masonry buildings, aesthetics, innovation, crisis

Background

AESTHETICS

Aesthetics are defined by the *Merriam-Webster Dictionary* as, "A set of ideas or opinions about beauty or art. A branch of philosophy dealing with the nature of beauty, art, and taste and with the creation and appreciation of beauty."

Aesthetics are subjective. What is considered beautiful by some people may not be considered beautiful by other people. For the purposes of this paper, the author's definition of aesthetics related to buildings with brick masonry exterior walls is *a building that is beautiful at the time it was constructed and that remained beautiful in future years after its construction.* An example of a building with brick masonry walls that is aesthetically beautiful according to the author's definition is Frank Lloyd Wright's Robie House, which was constructed circa 1910 in Chicago, IL (**Fig. 1** and **Fig. 2**).

FIG. 1 Robie House, circa 1910 (U. S. National Historic Landmark, U. S. National Register of Historic Places, and Chicago Landmark), designed by Frank Lloyd Wright. The cantilevered roof eaves and the low, long brick masonry walls with thick limestone coping create a strong and aesthetically pleasing horizontal expression of the building. The exterior brick masonry walls help to accentuate the aesthetically pleasing horizontal expression of the building.

FIG. 2 The exterior brick masonry walls include a veneer of red Roman brick that are 2 in. (50 mm) high by 12 in. (305 mm) long, with small head joints filled flush with colored mortar that matches the color of the brick and with tooled concave cream-colored mortar in 3/8 in. (10 mm) tall bed joints.

GREAT CHICAGO FIRE

The Great Chicago Fire that began on the evening of October 8, 1871 and lasted for almost three days affected a territory (burnt district) about 4 miles (6.5 km) in length by an average of two-thirds mile (1 km) in breadth (**Fig. 3**) [1,2]. The fire destroyed about 18,000 buildings [3].

At the time of the fire, an architect named John M. Van Osdel was the construction superintendent for a building within the burnt district. "As the fire approached, he gathered his plans and record books, took them to the basement, dug a pit, and placed the documents in it. He covered the pit with two feet of sand and a thick layer of damp clay. The records survived the fire, and suggested a method of fire-proofing with clay tile ..." [3, p.11].

Buildings Constructed within the Burnt District after the Fire (1871 to 2017)

Pertinent information obtained from an evaluation of buildings that were constructed with brick masonry exterior walls in the burnt district from 1871 to 2017, or approximately 150 years after the fire, was used to evaluate the topic of this paper: *Brick Masonry in Chicago: Crisis in Aesthetics?* The results of this evaluation are pertinent to buildings constructed in the burnt district during this period and may not apply to buildings constructed with brick masonry exterior walls elsewhere during this same time period.

FIG. 3 Chicago fire of 1871 burnt district.

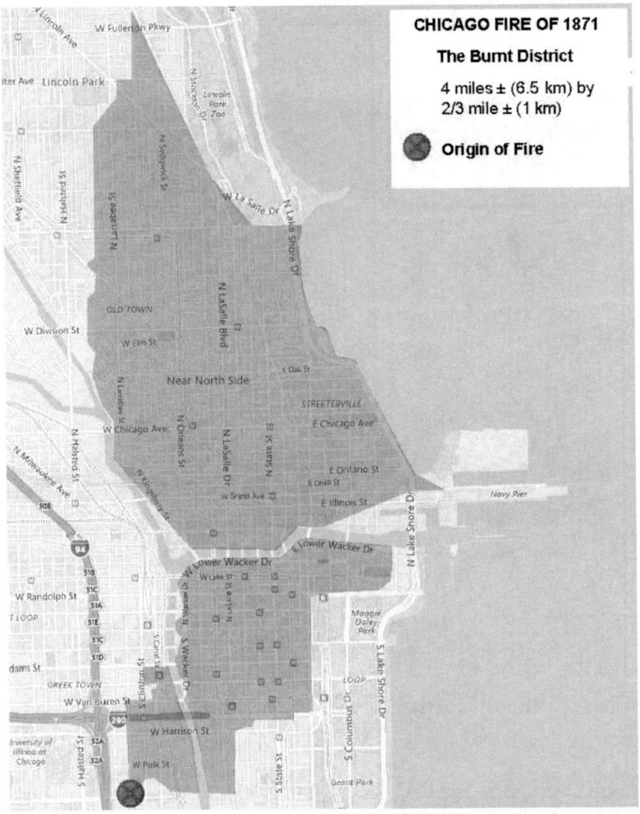

1871 TO 1930 (60 ± YEARS AFTER THE FIRE)

During this period, great advances were made in building construction that included:

1. In 1872, the Chicago City Council "passed an ordinance outlawing wooden buildings in the downtown area. Fireproofing was still in a primitive stage, and enforcement was never very effective; but this legislation requiring future construction to be of brick and iron did offer some protection against a grisly encore" [1, p. 118].
2. The first "fireproof" building was constructed in 1873, and more thought was being given to means of preventing another disaster by fire [3].
3. Elevators came into use during this period, allowing the higher buildings that flourished later [3].
4. The term "skyscraper" was first given to the 10-story Montauk Building (1882 to 1902) [3, p. 98].
5. The first building using skeleton construction, the Home Insurance Building, followed in 1885 [3, p. 98]. (Skeleton frame construction refers to a building in which all loads are transmitted to the foundation system by a skeleton of interconnecting columns and beams.)

6. Structural steel came into use, and wrought-iron beams and cast-iron columns began to be replaced [3, p. 98].
7. The fireproofing of columns and floor steel became standard procedure [3, p. 98].
8. Elevators and skeleton construction now allowed buildings to rise to heights of 12, 13, and 16 stories and then to the "21-story height of the wonder of the world, the Masonic Temple (1892)" [3, p. 98].

The exterior walls on many of the buildings that were constructed in the burnt district during this period included brick masonry. Several of these buildings utilized exterior walls of brick masonry in innovative forms to create aesthetically pleasing buildings. These included the Rookery Building, circa 1888 (**Fig. 4** and **Fig. 5**); the Marquette Building, circa 1895 (**Fig. 6** and **Fig. 7**); the Reid Murdoch Building, circa 1914 (**Fig. 8** and **Fig. 9**); Moody Church, circa 1925 (**Fig. 10** and **Fig. 11**); and a residential building, circa 1914 (**Fig. 12**). However, several buildings with brick masonry exterior walls that were also constructed in the burnt district during this period are not aesthetically pleasing (**Fig. 13**).

FIG. 4 The Rookery Building, circa 1888, 209 S. LaSalle Street (U. S. National Historic Landmark, U. S. National Register of Historic Places, and Chicago Landmark) designed by Burnham and Root. The brick masonry column covers with rounded corners and spandrels express the skeleton frame of this 12-story building. The brick masonry walls include a veneer of red brick set in dark-colored mortar in running bond. The combination of the semicircular brick arches above the seventh and tenth floor windows and the brick column covers with rounded corners create a powerful aesthetic expression of the building structure and of the blend of masonry wall systems.

FIG. 5 Masonry wall systems on the Rookery Building.

FIG. 6 The Marquette Building, circa 1895, 140 S. Dearborn Street (U. S. National Historic Landmark, U. S. National Register of Historic Places, and Chicago Landmark) designed by Holabird and Roche. The brick masonry and terra-cotta column covers and spandrels express the skeleton frame of this 16-story building. The combination of brick and terra-cotta creates a successful aesthetic expression of the building structure and blend of masonry materials and wall systems.

FIG. 7 Quoins in the brick masonry and between the terra-cotta panels on the Marquette Building. The building's exterior brick masonry walls include a veneer of dark brown brick set in running bond with dark-colored mortar and with an approximately 2.75 in. (70 mm) tall by 2 in. (50 mm) deep quoin every fourth brick course at the third through the fourteenth floors. On the first and second floors, reflected dark brown terra-cotta column covers with quoins between courses of terra-cotta are similar to the quoins in the brick masonry on the third through fourteenth floors.

FIG. 8 The Reid Murdoch Building, circa 1913, 320 N. Clark Street (U. S. National Register of Historic Places, and Chicago Landmark), designed by George C. Nimmons. The brick masonry column covers and spandrels, which were used to express the skeleton frame of this seven-story building, are capped with segmental brick arches above the seventh floor windows and a central brick-clad clock tower on the south elevation.

FIG. 9 Raked bed joints in a brick masonry wall on the Reid Murdoch Building. The exterior brick masonry walls include a veneer of red brick set in tan-colored mortar in running bond with a brick header course every seventh course. The flush mortar-filled head joints and the raked bed joints in the brick veneer accentuate the horizontal expression of the building that, when combined with the brick-clad skeleton frame, the brick segmental arches above the windows, and the brick-clad clock tower, create an aesthetically pleasing combination of brick masonry forms and systems.

1930 TO 2017 (60 ± TO 150 ± YEARS AFTER THE FIRE)

Very few buildings were constructed in the central business district section of the burnt district in the 1930s and 1940s [3]. This was due to the adverse effects of the financial crisis of 1929 and World War II (1939–1945).

Building construction increased from the 1950s through the 1980s and peaked around 1990 [3]. In 1974, the then-world's tallest building, the 110-story Sears Tower, was constructed.

Buildings with Exterior Walls of Glass/Aluminum Framed Curtain Walls, Concrete, or Stone

The majority of the significant buildings constructed in the burnt district during this period had exterior walls of glass/aluminum framed curtain walls, concrete, or stone. Some of these buildings that are aesthetically pleasing are the 33 W. Wacker

FIG. 10 Moody Church, circa 1925, 1634 N. LaSalle Street, designed by Fugard and Knapp. The building's exterior masonry walls were constructed with red brick set in cream-colored mortar in a running bond. There are dozens of semicircular arches in the building's exterior wall that include individual arches above the windows, the pairing of two arched windows within a larger arch (arcade opening), and arches above blind openings (blind arcades). The combination of these brick masonry arches creates an aesthetically pleasing expression of brick masonry application and form.

FIG. 11 Brick masonry arches on Moody Church.

FIG. 12 A residential building, circa 1914, 172 W. Eugenie Street. The exterior masonry walls of this three-story building include a veneer of red brick laid in running bond with tan-colored mortar. Some of the exterior walls form a bay window with a semicircular arch above the third-floor window below a steep gable roof and a segmental arch above the second-floor window with a projected water diverter. The bay window, the arches, and the steep gable roof create an aesthetically pleasing combination of brick masonry application and architectural forms.

FIG. 13 Office building, circa 1929.

Drive building, circa 1983 (**Fig. 14**); Marina City, circa 1968 (**Fig. 15**); and the Chase Tower (formerly the First National Bank building), circa 1968 (**Fig. 16**).

Buildings with Exterior Walls of Brick Masonry

During this period, no significant building with innovative, exciting, and aesthetically pleasing applications of brick masonry was constructed in the burnt district. Reasons why include:

1. The effects of World War I (1914–1918), the financial crash of 1929, and World War II (1939–1945) significantly slowed down construction of new buildings for approximately 30 years.
2. The number of architects and brick masons who had designed and constructed the aesthetic brick masonry clad buildings during the period approximately 60 years after the fire lessened considerably during this 30-year period of slower construction.
3. The cost of brick masonry wall systems increased due to:
   a. The weight and thickness of masonry walls increased the cost of the structural systems of the building.

FIG. 14 Glass/aluminum curtain wall, circa 1983 (33 W. Wacker Drive) designed by Kohn Pedersen Fox Associates. The design of this 36-story building was influenced by the adjacent Chicago River. The profile of the curved glass/aluminum curtain wall on the elevation that faces the river mimics the shape of the river, the blue/green glass in the curtain wall mimics the color of the water in the river, and the continuous horizontal bullnose-shaped mullions of the curtain wall mimics the flow of the water in the river. These conditions create the aesthetic effect of the building.

FIG. 15 Concrete façade, circa 1968, (Marina City, 300 N. State Street) designed by Bertrand Goldberg. Marina City includes two 64-story towers. The towers' structure is made of reinforced concrete. Each tower includes 40 floors of residential units above 18 ramped floors of parking. Each residential floor has 16 pie-shaped sections with flared, cantilevered, petal-shaped balconies that give the building its unique aesthetically pleasing "corn cob" appearance.

FIG. 16 Stone exterior wall, circa 1969, Chase Tower, formerly the First National Bank building, 21 S. Clark Street) designed by C. F. Murphy Associates. The exterior columns, spandrels, and walls of this 60-story building are clad with granite panels. The north and south facades have an inward sweeping curve that provides the large floor plate needed for banking transactions on the first floor and the smaller floor plates on the upper floors for office functions. The inward sweeping curve of the facades create an elegant aesthetically pleasing form that reflects the functional requirements of the building.

b. These walls, which consisted of many thousands of individual brick units with mortar, were constructed by hand.
c. The logistics of delivering masonry materials to locations in the buildings where the work was being performed also increased cost.
4. The owners and architects of proposed new buildings were uninspired.

Some of the buildings with exterior brick masonry walls that were constructed in the burnt district during this period utilized a combination of different colored brick in an attempt to make the buildings aesthetically appearling (**Fig. 17**). The aesthetics of this type of brick masonry application are not as innovative, exciting, and pleasing to look at as were those of the buildings with brick masonry exterior walls that were constructed in the burnt district approximately 60 years after the fire or of the buildings with glass/aluminum framed curtainwalls, metal panels, and stone that were constructed 60 to 150 years after.

AESTHETIC BUILDINGS WITH EXTERIOR WALLS OF BRICK MASONRY AROUND THE WORLD

Meanwhile, around the world, aesthetic buildings with brick masonry walls were designed and constructed. Various brick masonry designs were combined to create buildings that are exciting and pleasing to look at. These buildings include:

1. Musée Yves Saint Laurent in Marrakesh, Morocco, designed by Studio KO, Paris, France [4]. Construction was completed in 2017. The building has approximately 43,000 ft^2 (4,000 m^2) of floor area, of which approximately 4,300 ft^2 (400 m^2) is a permanent exhibition space devoted to the work of Yves Saint Laurent. "The facade of the building appears as an intersection of cubes with a lace-like covering of bricks, creating patterns that recall the weft and warp of fabric" [4].

FIG. 17 Buildings with brick masonry walls of different color brick.

2. MU:M Office Building, Munbal-dong, Paju-si, Gyeonggi-do, Republic of Korea, designed by Wise Architecture, Seoul [5]. Construction of the building was completed in 2015. "Entirely covered with black brick, the building has a twisted rectangular shape, and its entrance defined by the incised surface of the distorted structure..." [5].
3. Fundacion Santa Fe Hospital, Bogota, Colombia, designed by Giancarlo Mazzanti, Bogota [6]. Construction of the building was completed in 2016. "The Mazzanti team developed an ingenious system of thin, hollow bricks clipped onto cables like beads on a necklace, to screen the steel-to-concrete building, as well as to provide sun shading and, through a variety of apertures, frame views. The brick weave—loose in some areas, compact in others—creates varying degrees of transparency and opacity. A large concave depression and ribbon windows incise each façade, further adding texture to the four exposures" [6].

Conclusions

Based upon the evaluation of buildings constructed with brick masonry exterior walls in the Chicago Fire burnt district from 1871 to 2017, it is this author's opinion that there is an aesthetic crisis in significant buildings and that improvements should be considered. This opinion does not apply to buildings constructed with brick masonry exterior walls elsewhere during this same period.

The reasons why no significant aesthetically pleasing buildings with exterior brick masonry walls were constructed in the burnt district for approximately 60 to 150 years after the fire include the following:

1. The effects of World War I (1914–1918), the financial crash of 1929, and World War II (1939–1945) significantly slowed down construction of new buildings for approximately 30 years.
2. There were fewer architects and brick mason craftsmen during this 30-year construction slowdown than during the period of approximately 60 years after the fire.
3. The cost of brick masonry wall systems increased.
4. Owners and architects of proposed new buildings were uninspired.

Recommendations

It is important to encourage and educate building owners and architects on the aesthetic potential of brick masonry and its utilization in their buildings as follows.

ENCOURAGE AND EDUCATE ARCHITECTS AND OWNERS AS TO THE AESTHETIC POTENTIAL OF BRICK MASONRY

1. Pertinent design-related publications such as *Architectural Record* and *Architect* and online design magazines that include *Dezeen.com* and *Archdaily.com*, which are read by architects, owners, and developers of commercial, residential, and institutional buildings, regularly highlight the following:
   a. As-built aesthetic brick masonry buildings constructed in the United States and throughout the world

   b. Historic and modern details used in aesthetically pleasing brick masonry walls (arches, bond patterns, corbels, brick sizes and shapes, quoins, and so on)
   c. Examples of brick masonry durability
   d. Green properties of brick and other clay masonry products
2. The American Institute of Architects, the Brick Industry Association (BIA), and the International Masonry Institute should hold workshops for architects and owners on the aesthetically pleasing design of brick masonry buildings. In addition to focusing on the technical design aspects of brick masonry, these workshops should also focus on its aesthetic design aspects.
3. The American Institute of Architects, the BIA, and the International Masonry Institute should promote education about the aesthetic design of brick masonry buildings at architectural schools. Brick masonry design and construction courses are rarely taught in these schools, or the courses focus more on technical than on aesthetic design aspects of brick masonry.

GENERATE EXCITEMENT ABOUT THE AESTHETIC POTENTIAL OF BRICK MASONRY

1. Hold design competitions for buildings that focus on aesthetic or innovative use of brick masonry in buildings.
   a. Make the results of these competitions worthwhile for architects as well as owners.
   b. Publish invitations to and results of design competitions in journals and other publications and in online design magazines that are read by architects, owners, and developers of commercial, residential, and institutional buildings.
2. Since 1989, the BIA has sponsored the Brick in Architecture Awards, which are among the most prestigious awards programs featuring clay brick in the United States. The Best in Class award winners receive national recognition through a special Brick in Architecture insert in the December edition of *Architect* and in the BIA's *Brick News Online*.

The Brick in Architecture Awards highlight the best work in clay face and paving brick in the following categories: commercial, education, health care facilities, municipal/government, houses of worship, residential, paving and landscape, and renovation/restoration.

To further increase awareness of the aesthetic potential of brick masonry in buildings, the addition of an "innovation" category to showcase innovative use of brick masonry should be added to the Brick in Architecture Awards program.

References

[1] Mayer, H. M. and Wade, R. C., *Chicago: Growth of a Metropolis*, University of Chicago Press, Chicago, IL, 1969.

[2] *Encyclopaedia Britannica*, "Chicago Fire of 1871," https://web.archive.org/web/20171211191303/https://www.britannica.com/event/Chicago-fire-of-1871 (accessed November 12, 2017).

[3] Randall, F. A., *History of the Development of Building Construction in Chicago*, 2nd ed., University of Illinois Press, Urbana and Chicago, 1999.

[4] https://web.archive.org/web/20171211190702/https://www.museeyslmarrakech.com/en/ (accessed November 12, 2017).

[5] "MU:M Office Building/Wise Architecture," https://web.archive.org/web/20180807144157/https:/www.archdaily.com/800292/mu-m-office-building-wise-architecture (accessed August 7, 2018).

[6] Fixsen, A., "Secret Garden," *Architectural Record*, Vol. 205, No. 7, 2017, pp. 120–124.

STP 1612, 2018 / available online at www.astm.org / doi: 10.1520/STP161220170155

John M. Civitillo, Jr.[1] and J. Eric Peterson[2]

The Evolution of Stone Masonry Veneer Requirements: Past, Present, and Future

Citation

Civitillo, J. M. and Peterson, J. E., "The Evolution of Stone Masonry Veneer Requirements: Past, Present, and Future," *Masonry 2018, ASTM STP1612*, N. V. Krogstad and W. M. McGinley, Eds., ASTM International, West Conshohocken, PA, 2018, pp. 123–130, http://dx.doi.org/10.1520/STP161220170155[3]

ABSTRACT

Stone masonry has been one of the most durable and reliable building materials for thousands of years. Numerous stone structures still stand after centuries of dependable service and resistance to the elements. It has not been until recent history (the last 50 years) that the design and construction of masonry structures has been codified by building officials to impose minimum requirements for the performance of such construction. However, the evolution of modern building codes has resulted in a narrowing of options for constructing stone masonry veneer walls, and the current codes impose significant limitations in the prescriptive requirements for stone masonry veneer, which are often in conflict with industry recommendations and common construction practices. This forces many designers to rely solely upon engineering judgment for the design and construction of many commonly accepted stone masonry veneer wall systems. This paper explores the progression of code requirements for stone masonry from different, and sometimes conflicting, code bodies in the United States through the consolidation of the codes in the current International Building Code. It will also explore the suggested current direction of the masonry industry for expanding the code to include new allowable types, materials, and configurations

Manuscript received September 13, 2017; accepted for publication December 15, 2017.

[1]WDP Consulting Engineers, P.C., 53 W 36th St., Ste. 305, New York, NY 10018, USA http://orcid.org/0000-0002-9724-7477

[2]WDP & Associates Consulting Engineers, Inc., 10621 Gateway Blvd., Ste. 200, Manassas, VA 20110, USA

[3]ASTM International Symposium on *Masonry 2018* on June 26, 2018 in San Diego, CA, USA.

of stone masonry veneer systems to provide codified requirements for designers to utilize stone masonry more frequently in the design of buildings and other structures.

Keywords
stone, veneer, cavity

Introduction

Stone masonry construction, when properly designed and well-constructed, is renowned for its durability and resiliency. Many of the oldest surviving structures in the world are constructed of stone masonry materials.

The construction of masonry structures has changed significantly over the centuries. A number of structural and material improvements have been made, first with the addition of lime putty and sand mortar used to bond the interlocking masonry units, followed by the inclusion of Portland cement into the mortar composition in the late nineteenth century. The addition of reinforcing steel drastically improved the ability of masonry to accommodate internal tension and to resist imposed moments without the need to maintain conservative height to thickness ratios. However, for all the improvements in strength through the centuries, masonry wall systems were known to feel cold and damp, a function of the inherent thermal conductivity and mediocre water penetration resistance of early mass masonry construction.

Historically, stone masonry walls were constructed as mass masonry consisting of several wythes, or layers, of masonry bonded by interlocking masonry units laid within the wythes. Void spaces between masonry units were filled with rubble or filled solid with mortar. The effectiveness of a mass masonry wall to manage water is dependent solely on its moisture storage capacity and subsequent drying potential. Mass wall construction relies on the thickness of the wall to provide adequate storage for moisture permeating the exterior surface during a rain event and to allow the moisture to evaporate out of the wall before bulk water can reach the interior.

Through the years, masons experimented with variations of the mass masonry system (normally brick masonry) by incorporating air cavities or straw to provide improved insulation and to reduce the direct transfer of heat and cold through the wall. Cavity walls have been observed dating back to the Roman and Greek era, but the advantage of the water infiltration resistance was not documented until a British publication in the middle of the nineteenth century [1]. The concept was not adopted by a building or construction agency in the United States until 1937 [1]. In the early nineteenth century, true hollow walls, rowlock walls, and "rat trap" bond walls were utilized for brick masonry construction to connect the outer wythe of masonry across a partial cavity using bonded header courses. However, constructing stone masonry systems in this manner required significant effort to work the units into proper sizes to create such construction. This resulted in very few advances for stone masonry until the use of metal ties between wythes were employed to create multiwythe or composite construction masonry walls. While these construction methods were not often utilized for stone masonry, they set the stage for advancement of stone veneer construction by allowing integration of an irregular stone cladding with a more solid and consistent dimensioned back-up system. As this concept was advanced further in the middle of the twentieth century, a shift from mass masonry and composite wall construction to cavity wall or veneer wall construction began.

Masonry cavity wall construction refers to an assembly where multiple wythes of masonry are erected adjacent to each other with an air space between; but instead of filling the cavity solid between the wythes, the cavity is intentionally left open and often partially filled with insulating or weather-resistive materials (Fig. 1). The wythes are still tied together with discrete wall ties to allow the wythes to act compositely or simply resist out of plane loads, but the vertical load is accommodated independently within both the interior and exterior wythes. Cavity wall construction has many advantages over mass wall construction including improved thermal properties and reduced sound transmission; but the primary advantage of the cavity space is to provide a drainage plane that allows water to be managed internally and then evacuated from the wall system before reaching the inner wythe(s).

A cavity wall with a nonstructural outer wythe that provides aesthetic or weather-resistant capabilities is called a veneer. The veneer wythe "is not considered to add strength or stiffness to the wall system" [2]. Out-of-plane loading imparted

FIG. 1 Anchored stone masonry veneer cavity construction.

on the veneer is resisted by the structural backup wall through lateral connections between the wythes, and the veneer dead weight is supported directly by footings or intermediate shelf angles attached to the building structure. In anchored veneer, this lateral connection normally consists of a series of metal ties at a regular spacing between the two wythes, while adhered veneer is directly bonded to the backup and typically does not include a true cavity or air space.

Development and Evolution of Modern Building Codes

At the inception of the United States as a country and after the ratification of the Bill of Rights, there was a reluctance to permit the federal government to control building design and construction, thus regulation was left up to individual states and localities [3]. As natural disasters—such as fires, hurricanes, and earthquakes—devastated certain cities, awareness of the importance of regulating building construction to withstand such events for public safety became apparent. Coupled with a similar push from insurance companies to protect their interests, several privatized code agencies began to form [3].

The adoption of these codes by local governments was heavily weighted by geography, and the sometimes conflicting nature of these code bodies became apparent because each code body emphasized requirements that addressed the predominant environmental concerns (earthquakes, hurricanes, snow, and so on) of the associated locality. Building Officials and Code Administrators International, Inc. (BOCA) was established in 1915 and developed the Basic/National Building Code that was adopted by many principalities in the eastern United States. The Uniform Building Code (UBC) was used primarily in the central and western United States since its inception in 1927, and in 1940, the Standard Building Code (SBC) arose, which was enforced throughout the southern states.

Prior to 1977, there were multiple sources governing design and construction of masonry structures. Accordingly, the Masonry Standards Joint Committee (MSJC) was formed by the American Concrete Institute (ACI) and the American Society of Civil Engineers (ASCE) in 1977 to develop a central standard for the design of masonry structures; and the final approval of the first edition was achieved in 1989 [4]. The SBC and BOCA elected to adopt the joint masonry code ACI 530/ASCE 5/The Masonry Society (TMS) 402 [5,6], and their masonry provisions converged accordingly. However, the UBC maintained masonry design provisions internally [7].

Around the same time that the new joint masonry code began to be adopted by model code organizations, there was a clear understanding of the need for uniformity across the different building codes. ASCE Policy 340 ("Uniformity of Building Codes") was developed in 1989, and the policy made the argument for consistency among the various codes and their referenced standards [3]. Finally, in 1994, the three code groups agreed to merge and form the International Code Council;

and by 2000, the first International Building Code (IBC) was issued as a single, central building code body [8].

Stone Veneer Requirements

The origins of today's requirements for stone masonry veneer found in the current IBC heavily reflect those requirements of the UBC that date back to the 1960s. The 1967 edition of the UBC distinguished among different types of masonry veneer, including stone veneer up to 5 in. thick and stone veneer up to 10 in. thick [9]. The UBC permitted stone veneer up to 5 in. thick to be anchored with an air space of at least 1 in. provided the ties were spot-bedded with mortar. Stone veneer up to 10 in. thick had to be constructed with a 1 in. wide collar joint filled with grout. The modern IBC does not differentiate among thicknesses of stone veneer; however, the requirements for stone veneer up to 10 in. thick remain practically identical to those of the 1967 UBC, including the tie tributary area limit of 2 ft^2, the configuration of No. 12 gage steel wire ties over wood studs or concrete/masonry backing, and the solid grout-filled collar joint provision [10].

Meanwhile, BOCA and SBC were silent on the configuration of a veneer cavity until the beginning of the 1990s [5,6,11–13] at which point they began to reference the MSJC for anchored veneer. Around the same time, these two codes began to adopt the specific UBC provisions for stone veneer anchorage but only those relating to stone masonry not exceeding 10 in. thick such that the only stone veneer permissible by either code was one that was fully grouted. When the three codes were combined into the IBC, the fully grouted stone masonry veneer collar joint requirements from UBC governed due to the seismic importance of those provisions and the existing consistency of these requirements across all three codes. Indeed, commentary to the 2000 IBC states:

> According to the first [anchorage] method, the specified tie spacings, in conjunction with the cement grout, are intended to achieve lateral support points of the veneer at adequate spacings. Where lateral ties are not installed at the required location, buckling of the masonry veneer under vertical dead loads is possible. It should be noted that the required 1 inch (25 mm) of mortar grout creates walls that are not considered cavity wall construction [8].

Meanwhile, the 2000 IBC references the 1999 edition of ACI 530 (MSJC) that prescriptively requires one to, "maintain a 1 in. (25.4 mm) minimum air space" [14] for masonry veneer anchored to wood, steel, masonry, and concrete backings. As the MSJC does not treat stone masonry veneer independently, the two codes present a conflict in requirements for stone veneers that still exists within the current code documents.

Given that the apparent intent for the 2000 IBC requiring a grouted collar joint behind stone masonry veneer is to prevent buckling under seismic or other lateral loads (including eccentric dead loads), the authors suggest that the IBC should be modified such that stone veneer cavities need not be filled with grout if another means of buckling resistance can be provided or in localities where anticipated lateral loading conditions are not significant enough to result in buckling of the veneer. The code-mandated configuration of the wire ties as seen in **Fig. 2** relies on the tensile capacity and embedment of the wire ties to resist outward lateral loads, while the resistance of the wires under compression is significantly lower due to the tendency to buckle. The compressive resistance of the grout sustains the veneer under inward lateral loads. It is reasonable that if wire materials bridging a non-grouted cavity have sufficient compressive capacity capable of resisting inward force without buckling, they should be permissible by such a revised code, provided the stone veneer system between points of support was rigid enough to sustain the internal bending moments between points of support. Several requirements for anchored veneer, including tie spacing, were developed empirically based on existing spacing of masonry headers [15]. Accordingly, further research is necessary to understand the behavior of stone veneer systems under service lateral loads. For the high wind and seismic load regions of the current IBC, the fully grouted collar joint

FIG. 2 Stone veneer with concrete backing, 2000 IBC Figure 1405.6 [7].

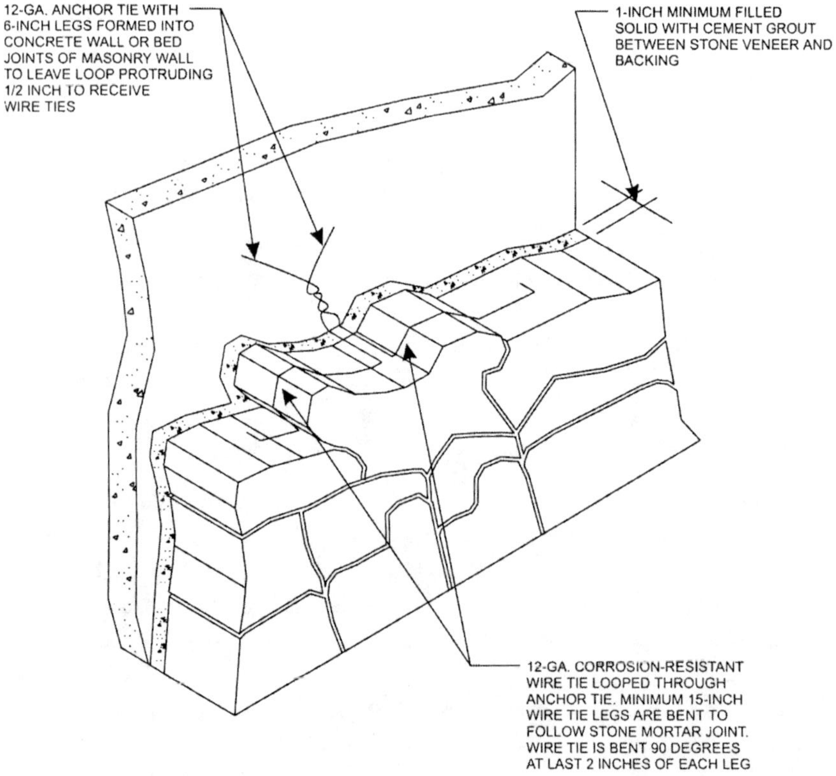

may still prove to be practically necessary in the proposed future IBC edition that permits stone veneer cavity air spaces when certain conditions are met if the required metal tie design and spacing is found to be uneconomical. However, for lower seismic and wind load regions, the code should permit the use of several commonly used cavity products that maintain drainage planes and air space along with cavity insulation.

It is important to note that the IBC, like BOCA and SBC before it, references the MSJC for anchored masonry [8,10]. The anchored masonry provisions of TMS 402 require a minimum 1 in. air space between the veneer and all backing materials [2,16]. Therein lies a direct conflict in the code that must be resolved because the veneer cavity cannot be fully grouted and simultaneously provide an air space.

Conclusions

It is well accepted in the industry that cavity wall construction reflects the ideal principles of a rain screen, which offers more dramatically improved weather and thermal performance than mass masonry walls. Further, the improved insulation configurations, fire resistance, and sound transmission render anchored veneer cavities a much more efficient and practical solution. However, the current limitation of today's IBC for stone masonry construction conflict with TMS 402 and do not permit the designer to take advantage of these benefits when designing with a stone veneer. Instead, the designer is limited to providing a solid collar joint or relying on alternative engineering justification rather than providing a rational design method. While this may be appropriate for areas with high wind or seismic considerations, which have been supported by the current IBC provisions that have been employed without modification since the late 1960s, a less restrictive alternative should be considered for regions with historically lower lateral load design conditions. Provisions could be developed for these areas that take into account the capacity of discrete, two-directional wire support points and the internal stress design of the veneer system to verify that the system is capable of resisting the design lateral loads while taking advantage of cavity or veneer wall construction. The provisions of the governing code bodies (both MSJC and IBC) should be updated to include these provisions to resolve the inherent conflict in requirements and to expand the ability of designers to employ systems utilizing stone masonry veneer.

References

[1] Satko, J., "History of Insulation with Masonry: Higher Energy Costs and Codes Demand Greater Efficiency," *The Masonry Edge*, Vol. 2, No. 2, n.d., pp. 32–41.

[2] Masonry Standards Joint Committee, "Building Code Requirements for Masonry Structures (TMS 402-12/ACI 530-13/ASCE 5-13) and Specifications for Masonry Structures (TMS 602-13/ACI 530.1-13/ASCE 6-13)," TMS, Longmont, CO, ACI, Farmington Hills, MI, ASCE, Reston, VA, 2013.

[3] Rossberg, J. and Leon, R. T., "Evolution of Codes in the USA," http://web.archive.org/web/20180808015925/https://www.nehrp.gov/pdf/UJNR_2013_Rossberg_Manuscript.pdf (accessed August 7, 2018).

[4] Brick Industry Association, "Technical Notes 3—Overview of Building Code Requirements for Masonry Structures (ACI 530-02/ASCE 5-02/TMS 402-02) and Specification for Masonry Structures (ACI 530.1-02/ASCE 6-02/TMS 602-02)," BIA, Reston, VA, 2002.

[5] Building Officials and Code Administrators International, "The BOCA National Building Code/1993," BOCA, Country Club Hills, IL, 1993.

[6] Southern Building Code Congress International, "Standard Building Code 1991 Edition," SBCCI, Birmingham, AL, 1991.

[7] International Conference of Building Officials, "1997 Uniform Building Code Volume I," ICBO, Whittier, CA, 1997.

[8] International Code Council, "2000 International Building Code," ICC, Falls Church, VA, 2000.

[9] International Conference of Building Officials, "Uniform Building Code 1967 Edition Volume I," ICBO, Pasadena, CA, 1967.

[10] International Code Council, "2015 International Building Code," ICC, Country Club Hills, IL, 2014.

[11] Building Officials Conference of America, "BOCA Basic Building Code," BOCA, Chicago, IL, 1965.

[12] Southern Building Code Congress International, "Standard Building Code 1976 Edition," SBCCI, Birmingham, AL, 1976.

[13] Building Officials and Code Administrators International, "The BOCA National Building Code/1990," BOCA, Country Club Hills, IL, 1989.

[14] Masonry Standards Joint Committee, "Building Code Requirements for Masonry Structures (ACI 530-99/ASCE 5-99/TMS 402-99) and Specifications for Masonry Structures (ACI 530.1-99/ASCE 6-99/TMS 602-99)," ACI, SEI, TMS, Farmington Hills, MI, Reston, VA, Boulder, CO, 1999.

[15] Borchelt, J. G., "History of Anchored Masonry Veneer," *Proceedings of the 8th International Brick Block Masonry Conference*, Dublin, Ireland, September 19–21, 1988, Elsevier Science Publishing Co., New York, pp.1496–1506.

[16] ACI-ASCE Committee 530, "Building Code Requirements for Masonry Structures (ACI 530-88/ASCE 5-88) and Specifications for Masonry Structures (ACI 530.1-88/ASCE 6-88)," ACI, New York, NY, 1989.

STP 1612, 2018 / available online at www.astm.org / doi: 10.1520/STP161220170184

Salah Sarhat[1] and Edward G. Sherwood[2]

Shear Strength of GFRP-Reinforced Concrete Masonry Beams

Citation

Sarhat, S. and Sherwood, E. G., "Shear Strength of GFRP-Reinforced Concrete Masonry Beams," *Masonry 2018, ASTM STP1612*, N. V. Krogstad and W. M. McGinley, Eds., ASTM International, West Conshohocken, PA, 2018, pp. 131–157, http://dx.doi.org/10.1520/STP161220170184[3]

ABSTRACT

This paper studies the effect of using glass fiber-reinforced polymer (GFRP) flexural reinforcement on the shear strength of reinforced masonry (RM) beams. For this purpose, two sets of shear-critical RM beams were tested, one set being two courses high, the other being five courses high. Each set consisted of two RM beams, one reinforced with GFRP and one reinforced with high-strength steel. The reinforcement ratio was kept constant for all beams. The test results showed that the use of GFRP reinforcement resulted in considerably reduced shear strength compared to beams reinforced with steel. Furthermore, the failure shear stresses of the beams decreased significantly due to increases in depth indicating that size effect is real in reinforced masonry. The shear behavior of the RM beams was seen to be fundamentally similar to that of reinforced concrete beams. As shear design expressions in most of the masonry design codes, such as TMS 402-2016 code, are intended to be used for designing masonry beams reinforced with conventional steel, the test results showed that these expressions can yield unsafe shear predictions for GFRP-reinforced or high-strength steel-reinforced masonry beams. The general method of the Canadian Standards Association (CSA) S304-2014 code, on the other hand, was able to capture safely the variations in the shear strength of GFRP-reinforced masonry. The general

Manuscript received September 30, 2017; accepted for publication December 27, 2017.

[1]NSERC CREATE Sustainable Engineering in Remote Areas (SERA) Program, Queen's University, Dept. of Civil Engineering, 58 University Ave., Kingston, ON, Canada http://orcid.org/0000-0002-3011-7379

[2]Carleton University, Dept. of Civil and Environmental Engineering, 1125 Colonel By Dr., Ottawa, ON, Canada K1S 5B6 http://orcid.org/0000-0002-6715-4288

[3]ASTM International Symposium on *Masonry 2018* on June 26, 2018 in San Diego, CA, USA.

method is based on the modified compression field theory and can account for all influential factors affecting the shear strength of reinforced masonry beams. The results of this paper highlight the necessity of revising masonry design codes to address the size effect and the effect of reinforcement stiffness.

Keywords
FRP, masonry design codes, reinforced masonry beams, size effect

Introduction

As reinforcing material properties improve and higher-strength reinforcing rebars, such as high-strength steel and fiber-reinforced polymer (FRP), are produced at competitive costs, it is expected that designers will utilize them to reinforce masonry elements. As such, masonry design codes need to be evaluated to check if they can effectively accommodate the use of these higher-strength reinforcing bars. Noncorrosive FRP reinforcing bars, in particular, have superior durability and a high strength-to-weight ratio, which makes them attractive as an internal reinforcement method for reinforced masonry (RM) elements such as beams [1]. RM beams are common masonry structures used to span various openings, such as doors, windows, and passages. Also, RM beams may be used as coupling beams connecting between masonry shear walls or piers [2]. As it is difficult to provide shear reinforcement in masonry, the design provisions used to determine shear strength of RM beams without web reinforcement must be accurate, safe, and rational [3].

The flexural behavior of masonry beams and walls reinforced with glass fiber-reinforced polymer (GFRP) was experimentally studied [1,4,5] and GFRP reinforcement was found to be effective and practical. The shear behavior of GFRP-reinforced masonry beams, to the authors' best knowledge, has not been studied yet. This is despite the fact that the shear behavior of GFRP-reinforced concrete beams has been thoroughly studied by numerous researchers [6–10]. As such, it is important to explore the effect of using GFRP reinforcing bars on the shear strength of RM beams.

In this investigation, two sets of RM beams were tested to study the effect of using GFRP reinforcing bars versus high-strength steel on the shear strength of RM beams. In each set, the depth and the reinforcement ratio were kept constant, and the only variable was the reinforcement type (GFRP versus steel). Furthermore, the results from the two sets of RM beams were used to study the effect of beam depth on the shear strength of both types of reinforcement. The experimental shear strengths were compared with the failure shear stresses predicted using two different North American masonry design codes: the CSA S304-2014 [11] and TMS 402–2016 [12] codes.

Shear Design Methods for RM Beams

It is widely recognized that RM beams exhibit many similarities in shear behavior to reinforced concrete beams [3]. The applied shear force (V_c) in reinforced concrete members without transverse reinforcement is resisted by three major components (as shown in Fig. 1): the force in the compression zone (V_{cz}), the vertical component of the aggregate interlock shear stress (v_a), and the dowel force (V_d) in the tensile steel. A large number of studies and investigations have been conducted over the years to determine the relative proportions of each of these components; however, these proportions still represent a source of debate. The main parameters that affect the shear strength of reinforced concrete beams are shear span-to-depth ratio, reinforcement ratio, effective depth (size), reinforcement stiffness and distribution, crack width and roughness, concrete strength, and cover thickness.

CSA S304-2014

The shear design provisions for RM beams in the newly released Canadian code (CSA S304-2014 [11]) are based on the simplified modified compression field theory (SMCFT) originally developed for reinforced concrete as described by Bentz, Vecchio, and Collins [13]. In the SMCFT, it is assumed that aggregate interlock, developed across cracks in the masonry below the neutral axis, is the main shear transfer mechanism in RM beams without web reinforcement. According to the SMCFT, the width of a crack, w, can be calculated as the product of the average strain normal to the crack, ε, and the spacing of the cracks, s, in the normal direction (i.e., $w = \varepsilon \times s$). Test results by Walraven [14] have shown that the aggregate interlock capacity of a crack is inversely related to the width of that crack. As such, it can be concluded that any action that increases the longitudinal strain in a member or increases the crack spacing will reduce the aggregate interlock and hence the shear strength because of the resulting wider cracks. The CSA S304-2014 code is the only masonry design code that accounts for the size effect, wherein the failure shear stress decreases as the effective depth increases due to reduced aggregate interlock capacity. Likewise, the CSA S304-2014 design provisions account for the strain effect, wherein the failure shear stresses decrease as longitudinal strains increase resulting in reduced the aggregate interlock.

The CSA S304-2014 general shear design method uses the following basic equation to calculate the shear strength of RM beams without stirrups:

$$V_m = K_b \beta \sqrt{f'_m} b_w d_v \tag{1}$$

FIG. 1 Mechanisms of shear transfer in reinforced concrete without stirrups.

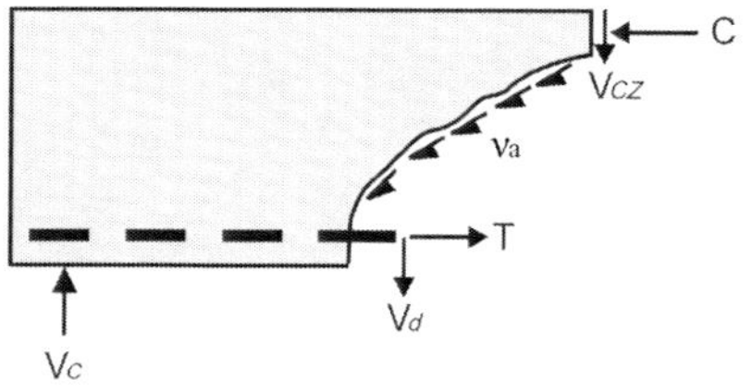

where:

b_w = overall web width of the beam (mm),

K_b = a factor that accounts for the type of masonry and is equal to 1.0 for grouted hollow masonry,

β = a factor that describes the ability of cracked masonry to transfer shear stress by aggregate interlock,

f'_m = masonry compressive strength,

d_v effective depth for shear calculations, taken as the greater of 0.9 d or 0.72 h, mm, and

h = overall height of the beam, mm.

The term β is calculated as the product of a strain effect term and a size effect term:

$$\beta = \frac{0.4}{(1 + 1500\,\varepsilon_x)} \cdot \frac{1300}{(1000 + z_e)} = (\textit{Strain effect term})(\textit{Size effect term}) \qquad (2)$$

For longitudinal strain at the mid-depth of RM sections that are neither prestressed nor subjected to axial loads, ε_x, is a function of (1) the factored applied moment, M_f and shear V_f at the section, and (2) the stiffness (E_s) and area (A_s) of the longitudinal flexural steel. The equation for calculating ε_x is as follow:

$$\varepsilon_x = \frac{M_f/d_v + V_f}{2E_sA_s} \qquad (3)$$

The term z_e is an "equivalent crack spacing factor" that models the effects of different maximum aggregate size (a_g):

$$z_e = g_a\, z_s \qquad (4)$$

where:

g_a = 1.4 for course grout and 1.7 for fine grout, and

z_s = the lesser of dv or the vertical spacing between layers of intermediate reinforcement.

To use the CSA S304 general method to calculate shear strengths of experimental specimens, Eqs 1 through 4 must be applied at the critical section by solving a quadratic equation or using an iterative approach [15]. To demonstrate the CSA S304 general method in use, consider a strength prediction of Specimen SM4 described later in this paper. For this specimen, d_v = 792 mm, b_w = 190 mm, A_s = 548 mm^2, f'_m = 18.5 MPa, E_s = 200 GPa, and the shear span is 2,750 mm. The crack spacing parameter, z_s, is equal to 792 mm. With fine grout, z_e in Eq 4 will be equal to 1,346 mm and the size effect term in Eq 2 will be equal to 0.554. Considering a critical section at d from point load, the moment due to self-weight is 13.73 kN.m. Assuming that $\varepsilon_x = 0.5 \times 10^{-3}$ will lead to an estimated shear failure load from Eq 1 of V_m = 282.2 kN. Calculating Eq 3 based on the first estimate of V_m produces a new estimate of $\varepsilon_x = 1.316 \times 10^{-3}$, indicating that the original estimate was not correct and more iterations are needed. Convergence is attained at a strain of $\varepsilon_x = 0.946 \times 10^{-3}$ and a corresponding shear force of 58.9 kN. Specimen SM4 failed in shear at 68.9 KN or at a test to predicted ratio of 1.17.

HOULT, SHERWOOD, BENTZ, AND COLLINS METHOD

Hoult, Sherwood, Bentz, and Collins [15] proposed a more accurate version of Eq 5 specifically intended for reinforced concrete members with high strains (reinforced with very low reinforcement ratios or with low stiffness reinforcement). The equation can be adapted to masonry as follows:

$$\beta = \frac{0.3}{0.5 + (0.15 + 1,000\,\varepsilon_x)^{0.7}} \cdot \frac{1300}{(1,000 + Z_e)} \tag{5}$$

TMS 402-2016

It should be noted that TMS 402 shear provisions restrict reinforcing bars to conventional steel reinforcing with yield strengths that do not exceed 60 ksi (414 MPa). TMS 402 shear expression is not intended to be used to design masonry members reinforced with high-strength steel or FRP rebars.

The nominal shear strength provided by masonry is computed using Eq 6 (in psi units):

$$V_m = [4.0 - 1.75(M_u/V_u d_v)]A_n\sqrt{f'_m} \tag{6}$$

where:

$M_u/V_u d_v = a/d$ for point-loaded beams, and need not be greater than 1.0, and

A_n = net cross-sectional area of beam (taken as $b_w\, h$).

For slender beams with $a/d > 1.0$, the formula can be simplified to Eq 7 (in MPa units):

$$v_m = V_m/b_w h = 0.187\sqrt{f'_m} \tag{7}$$

Experimental Program

Four RM beams without stirrups were constructed and loaded to failure to study the effect of using GFRP flexural reinforcement on the shear strength of RM beams. These beams represent one phase of an extensive experimental program that was recently completed at Carleton University [16] to investigate the shear strength of masonry beams. Multiple variables were studied as part of this program, including the size effect (effective depth) and the use of nonconventional reinforcement steel (high-strength steel or FRP rebars). Three notations were used in labeling the beams. The first notation is the letter *S*, which refers to the first letter of the last name of the author. The second notation refers to the size of the beam: *M* for five-course high beams and *S* for two-course high beams. The third notation is a number that refers to the type of internal reinforcement type: four for high-strength steel and five for GFRP. **Fig. 2** demonstrates the cross sections, reinforcement details, and instrumentations of the tested RM beams.

The four beams consist of two sets of full-scale shear-critical fully grouted RM beams. The first set consisted of two RM beams, namely SS4 and SS5. These beams were two courses high with Specimen SS4 reinforced with a 15-mm high-strength steel bar, and Specimen SS5 reinforced with a #5 GFRP bar. The second set consisted of two five-course-high RM beams: SM4 was reinforced with a 26-mm high-strength steel bar, and SM5 was reinforced with a #8 GFRP bar. Specimens SM4 and SM5 were additionally reinforced with two 20M conventional steel bars in the upper course as compression reinforcement. The *a*/*d* ratios for all beams were approximately constant at 3.13, ranging from to 3.10 to 3.16. **Table 1** summarizes the geometry, reinforcement details, and material properties of the tested RM beams. The mechanical properties of the high-strength steel (Dywidag rebars) and the GFRP rods were obtained from the manufacturers. The effect of using different reinforcement stiffnesses (steel rebars versus GFRP) on shear strength can be investigated by comparing the results of SS4 to SS5 and SM4 to SM5. The effect of beam depth, on the other hand, can be studied by comparing the results of SS4 to SM4 and SS5 to SM5.

All the beams and companion specimens were built in the laboratory by experienced masons under the supervision of the authors, and their quality of workmanship was rated to be very good. Lintel units were used in the first course, while the top courses were built with depressed web units (**Fig. 3a**). The construction sequence was as follows: first, the lower course was laid; next, the reinforcing rebars were placed; and then, the upper courses were laid with face shell bedding. The average block unit strength was 36.9 MPa (based on net area). The compression tests on block units were conducted in accordance with CSA A165.1-04 [17]. In order to

FIG. 2 Cross sections, reinforcement details, and instrumentations of the tested reinforced masonry beams.

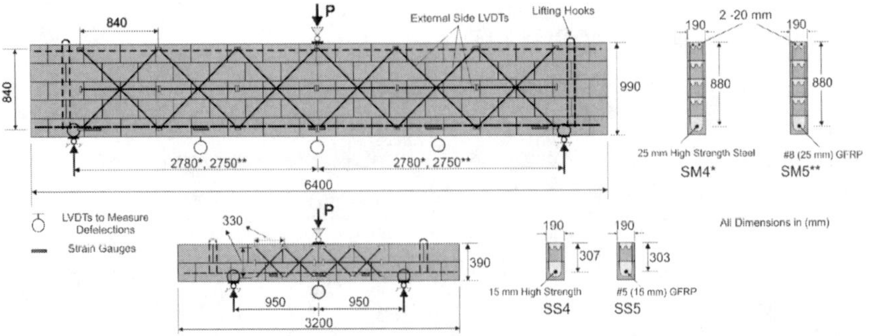

TABLE 1 Geometry, reinforcement details and material properties of the reinforced masonry beams.

	Geometry (As Built)						Reinforcement						Masonry Properties			
Beam ID	L (mm)	h (mm)	a (mm)	d (mm)	a/d	b_w (mm)	Steel Type	db (mm)	A_S (mm^2)	E (GPa)	Yield Strength f_y (MPa)	ρ (%)	f_{bl} (Net Area) (MPa)	f_{mr} (MPa)	f_{gr} (MPa)	f'_m (MPa)
SS4	3,200	390	950	307	3.10	190	Steel	15	177	205	900	0.30	37.5	25.5	19.6	18.0
SS5	3,200	390	950	303	3.14	190	# 5 GFRP	15.8	197	48.2	656	0.34	37.5	26.3	18.3	17.5
SM4	6,400	990	2,780	880	3.16	190	Steel	26	548	205	830	0.33	36.3	26.1	19.3	18.5
SM5	6,400	988	2,750	880	3.13	190	# 8 GFRP	25.4	507	51.9	675	0.30	36.3	27.4	20.5	19.5

obtain the compressive strength of the masonry beams, four-unit-high normal to bed joint prisms were constructed from the same material at the time of construction to establish f'_m. Furthermore, two prisms that were to be loaded parallel to the bed joint were built and tested to determine the modulus of elasticity of the masonry for Specimens SS5 and SM5 as discussed later. Type S mortar with an average compressive strength of 26 MPa was used in all beams. Mortar was batched, mixed, and tested for compressive strength in accordance with CSA A179-04 [18] and ASTM C109/C109M, *Standard Test Method for Compressive Strength of Hydraulic Cement Mortars (Using 2-in. or [50-mm] Cube Specimens)* [19]. The beams and the prisms were fully grouted with a fine grout mix with an average compressive strength of 19 MPa. The strength of the grout was determined in accordance with CSA A179-04 [18] and ASTM C1019, *Standard Test Method for Sampling and Testing Grout* [20].The grout was supplied by a local ready-mix company and had a slump of 220 mm. The grout was carefully vibrated, and the top surfaces of the specimens were smoothly leveled and covered with plastic sheeting for seven days at room temperature. **Fig. 3** shows photos of the construction process. The masonry prisms were capped using hydrostone plasters conforming to ASTM C1552, *Standard Practice for Capping Concrete Masonry Units, Related Units and Masonry Prisms for Compression Testing* [21], and were tested under concentric axial loading in accordance with SCA S304-14 [12], Annex D.

The main reinforcing rebars were instrumented with multiple strain gauges at different locations as indicated in **Fig. 2**. The specimens were tested in three-point bending under monotonic load until shear failure occurred. Vertical deflection was measured at different locations, and longitudinal, transverse, and shear strains were measured on the side faces of each specimen at the different locations in each shear span using side-mounted linear-variable displacement transducers as shown in **Fig. 2**. Loading was stopped regularly during the test at "load stages." At load stages, the load was reduced by 10 % for safety, cracks were marked and widths measured with crack comparators to a precision of 0.025 mm, and photos were taken.

Experimental Results

LOAD-DEFLECTION BEHAVIOR

All specimens exhibited a shear failure before reaching their flexural capacity. Shear stress-relative displacement curves are presented in **Fig. 4**. It can be seen in the figure that failures generally were brittle with the exception of Specimen SM5 as discussed later. No post-peak response was observed in any beam. Similar flexural cracking loads were observed for both steel- and GFRP-reinforced beams in each series. It can also be seen that post-cracking stiffness of GFRP-RM beams decreased significantly compared to their corresponding steel-reinforced beams. The major observation in the current study, however, is the fact that the maximum loads reached in GFRP-reinforced beams were significantly lower than their corresponding steel-reinforced beams. Another observation is that increasing the effective depths led to substantial reductions in failure shear stresses in both GFRP and steel RM beams.

FIG. 3 (a) Construction process of the beams, (b) grouting of the beams, (c) normal to bed joint prisms, and (d) parallel to bed joint prisms.

CRACKING BEHAVIOR AND FAILURE MECHANISMS

All specimens (except Specimen SM5) exhibited clear diagonal shear failures due to the breakdown of beam action before reaching their flexural capacity. The progression of shear failure in Specimen SS4 (reinforced with steel) is presented in Fig. 5. In this figure, it can be seen that flexural cracks appeared at early load stages

FIG. 4 Shear stress-relative displacement responses of Specimens SS4, SS5, SM4, and SM5.

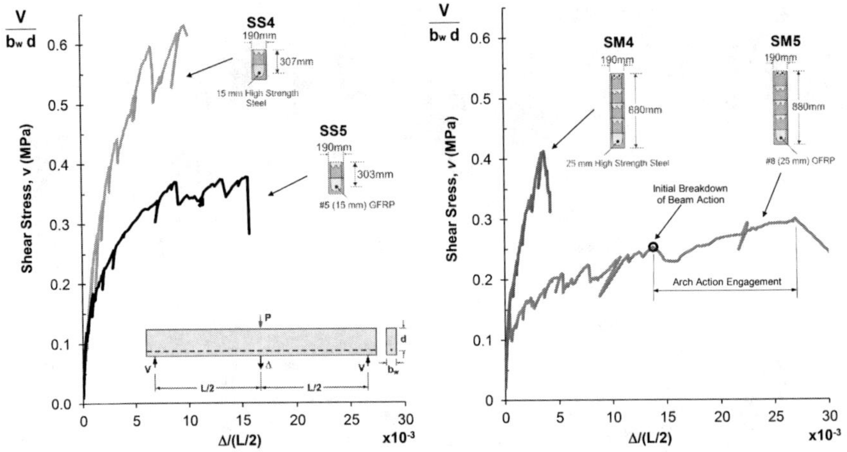

FIG. 5 Failure sequence of Specimen SS4.

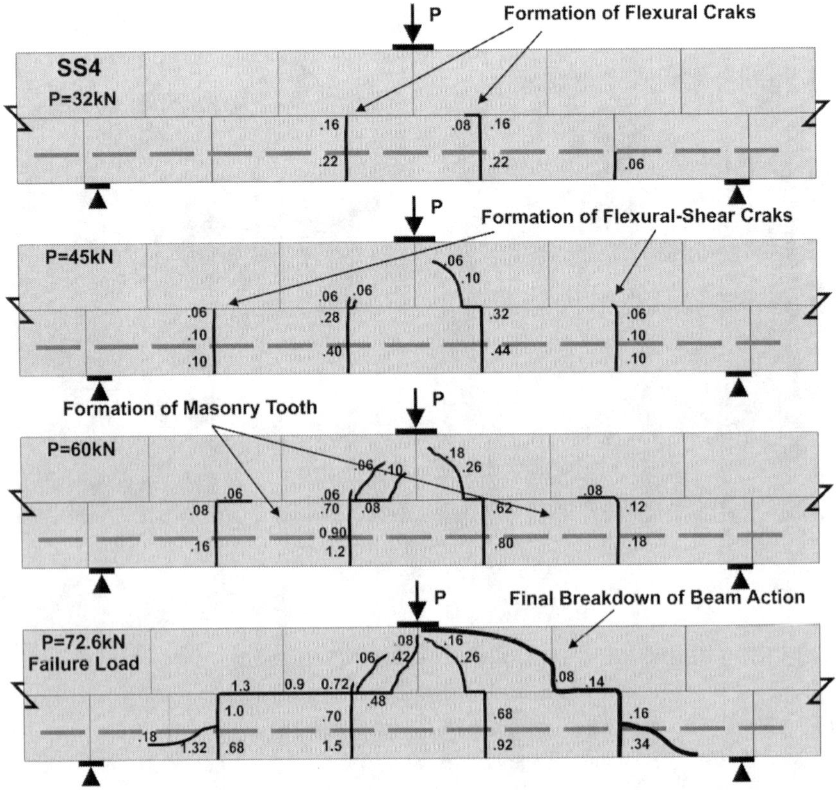

in the head joints of the bottom course. With more load, flexural-shear cracks appeared in both shear spans through the head joints far from the point load and extended up toward the second course in each span. Upon further loading, the flexural-shear cracks divided the tensile zone into parts, or "masonry teeth," isolated between two consecutive cracks in a similar fashion to reinforced concrete. At a load of 72.6 kN, the far crack in the west span (right side in Fig. 5) extended upward through the head joint and diagonally toward the point load causing a clear diagonal tension failure. Reflecting the extensive increase in the crack width, Fig. 6 shows that shear strains in the west span increased significantly indicating that shear stresses can no longer be transferred through aggregate interlock. Furthermore, it can also be noted that cracks in RM beams form at predefined paths (mortar joints), giving the cracks a more jagged appearance than in RC beams.

Fig. 7 shows that the progression of shear failure in Specimen SS5 (reinforced with #5 GFRP bar) was similar to that of SS4 (reinforced with steel). However, the jagged nature of the diagonal crack was more apparent in SS5 than in SS4, which means that the aggregate interlock occurred on different planes. It can be seen in Fig. 7 that the dominant inclined crack, which caused failure in the east side of the beam, passed through three regions: vertically through the head joint, horizontally

FIG. 6 Load versus shear strain in Specimen SS4.

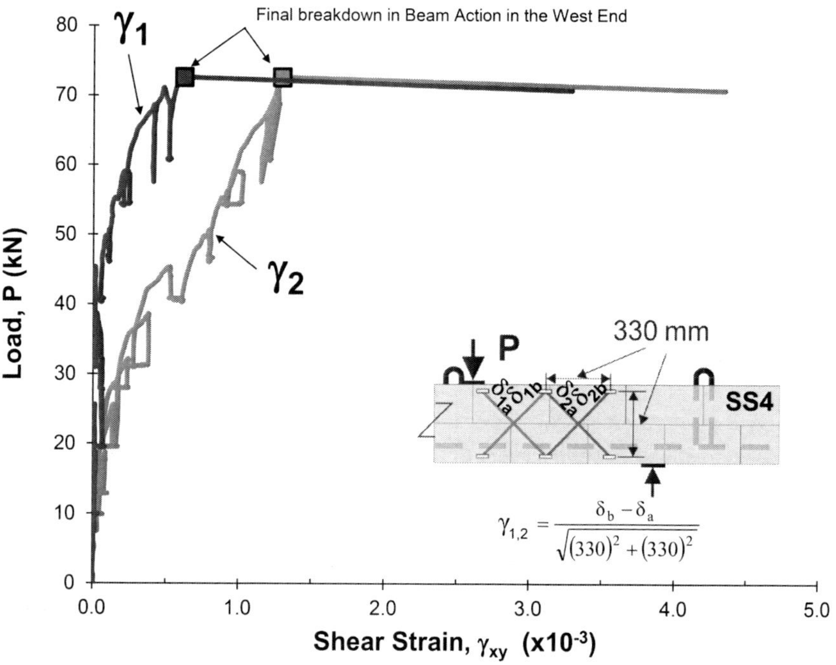

FIG. 7 Failure sequence of Specimen SS5.

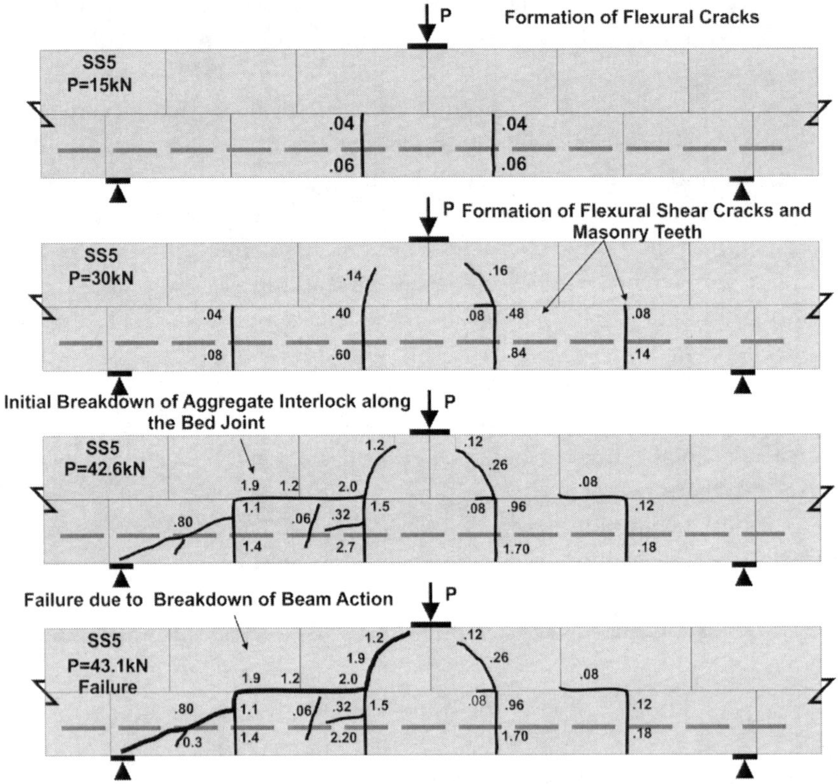

through the bed joint, and diagonally through the units. The shear strains measured in the west span show that the breakdown in the beam action in SS5 was gradual (**Fig. 8**). It seems that the aggregate interlock was initially lost along the bed joint, and this resulted in the formation of a shear key at corners of masonry units that permitted shear stresses to be resisted by the interlocking of adjacent blocks in addition to the aggregate interlock across the rest of diagonal jagged crack. This gradual breakdown of beam action can explain why the failure of SS5 was associated with larger deflection than the initial breakdown of aggregate interlock as shown in **Fig. 4**.

The progression of failure of Specimen SM4 is presented in **Fig. 9**. The cracking sequence of the five-course high beams was similar to those of two-course high beams. The failure of SM4 happened in a very similar manner to that of reinforced concrete beams as a diagonal tension crack suddenly formed at a load of 130.8kN in the east span causing failure. The crack propagated upward toward the point load and extended backward at the bottom bed joint toward the support. Specimen

FIG. 8 Load versus shear strain in Specimen SS5.

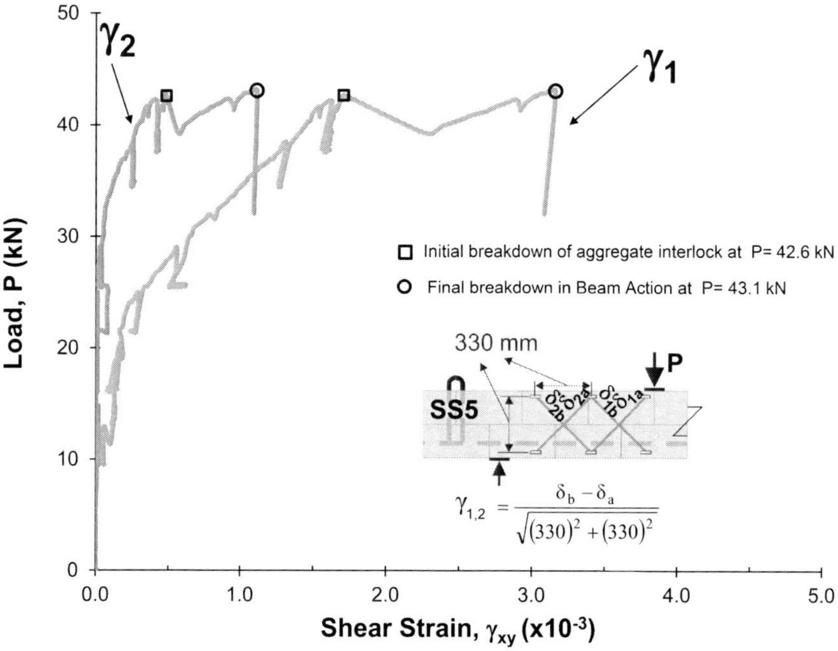

FIG. 9 Failure sequence of Specimen SM4.

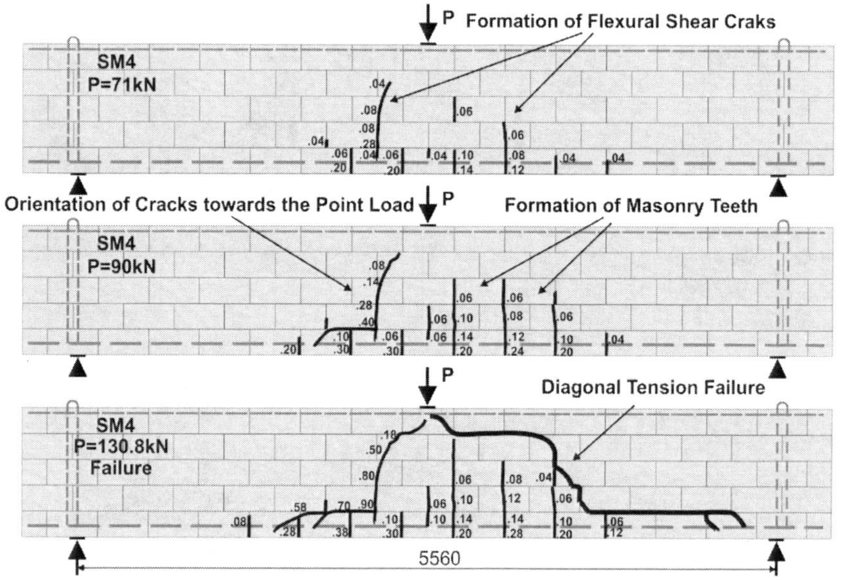

SM5, however, was characterized by a far more gradual mechanism. An initial breakdown of beam action at a load of 77 kN (equivalent to 80 % of the failure load) was followed by formation of shear keys between the two sides of the vertical part (detail A in **Fig. 10**) of the major diagonal crack in the east span. The videotape of the failure showed that these shear keys provided a substantial shear resistance mechanism. The gradual extension and widening of the crack along the lowest bed joint (detail B in **Fig. 10**) suggested that a shear resistance mechanism was also engaged along the bottom bed joint. It can be noted in **Fig. 10** that the compression zone above the major diagonal crack was very shallow (less than 20 mm). As such, the compression zone did not contribute much to resisting shear force across the section. The steel strain in the east support exhibited a considerable increase after the initial breakdown of the beam action. However, the steel strain at the support was less than 50 % of that at mid-span, indicating that strut and tie action was slightly engaged and that the majority of shear resistance occurred by the shear key formed at the vertical part of the diagonal crack. It is more likely that arch action was engaged as a result of the dominant crack being close to the loading point leaving a large uncracked portion of the beam in which an effective strut can form extending from the point load toward the support. Such behavior was also observed in Specimen R-G007Na tested by Tariq and Newhook [6] and in Specimen MB1 tested by Khaja and Sherwood [10]. **Fig. 11** and **Fig. 12** show the appearance of the beams after the tests were completed.

FIG. 10 Failure sequence of Specimen SM5.

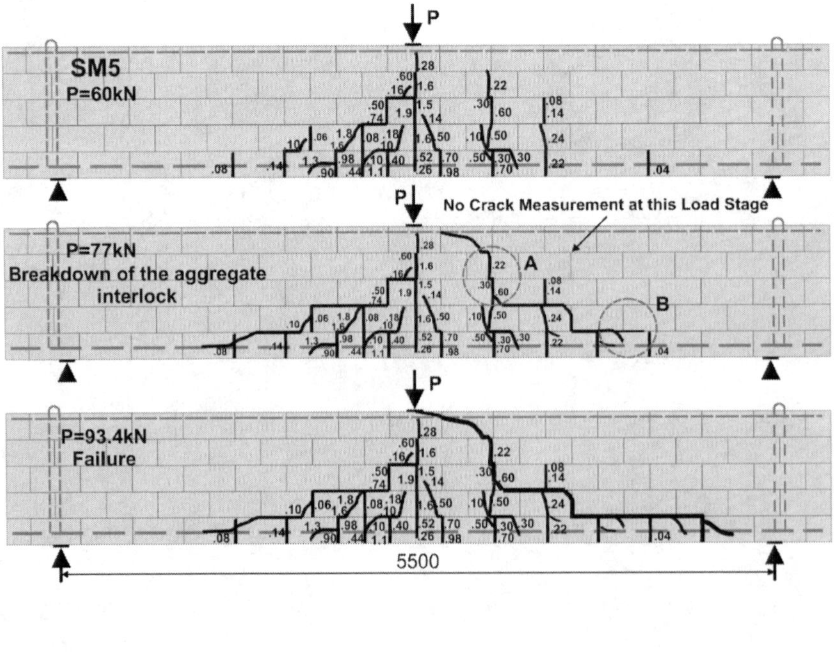

FIG. 11 The appearance of SS4 and SS5 beams after the tests were completed (cracks digitally enhanced for clarity).

SHEAR STRESS TRANSMITTED IN COMPRESSION ZONE

The shear design provisions in masonry design codes such as TMS 402 are empirically derived without explicit consideration of the role that interlocking below the neutral axis plays in shear resistance. This is despite the fact that several tests in reinforced concrete [22–25] have demonstrated that shear stresses transferred by aggregate interlock are significant. To experimentally quantify the contribution of the compression zone to the shear resistance in RM, two columns of strain gauges were affixed at the tip of an expected shear crack in the west span of Specimen SS5 (**Fig. 13**). The longitudinal masonry strains measured by these strain gauges were converted to longitudinal stresses at a load of 41 kN (equivalent to 95 % of ultimate shear strength) using a stress-strain curve constructed for a companion masonry prism tested under compression parallel to the bed joints following a similar load history as Beam SS5. The distribution of shear stresses in the compression zone can be constructed from the masonry flexural stress profiles by using the basic formula of ($v = VQ/Ib$) developed by Jourawski [26]. By integrating the resulting shear stress profile, it can be seen that the uncracked compression zone in Specimen SS5 carried only 26 % of the total shear applied by the load, leaving the majority of the vertical shear in SS5 to be resisted by aggregate interlock in the web of the beam (v_{agg}) and

FIG. 12 The appearance of SM4 and SM5 beams after the tests were completed (cracks digitally enhanced for clarity).

dowel action (V_d). Knowing that the dowel action resists only a small portion of the vertical shear [23,27,28], it can thus be concluded that the aggregate interlock is the main shear transfer mechanism in RM. It is worth mentioning that aggregate interlock can be associated with the formation of shear keys at corners of masonry units resulting from the jagged nature of shear cracks in masonry. These shear keys offer additional shear resistance by the interlocking of adjacent blocks, a phenomenon not generally observed in concrete.

A similar procedure was followed for Specimen SM5. **Fig. 14** shows the measurement of shear forces transmitted by the uncracked compression zone in

FIG. 13 Measurement of shear transfer portion by compression zone in Specimen SS5.

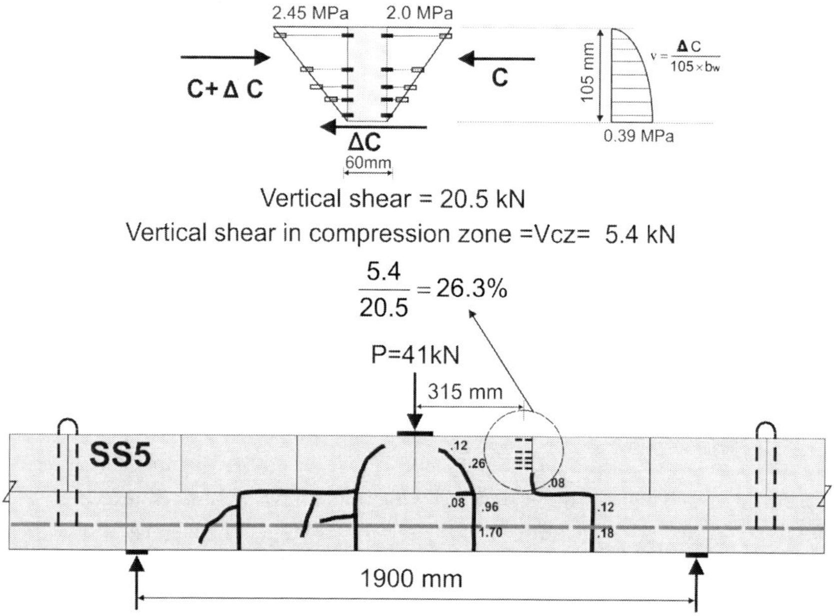

FIG. 14 Measurement of shear transfer portion by compression zone in Specimen SM5.

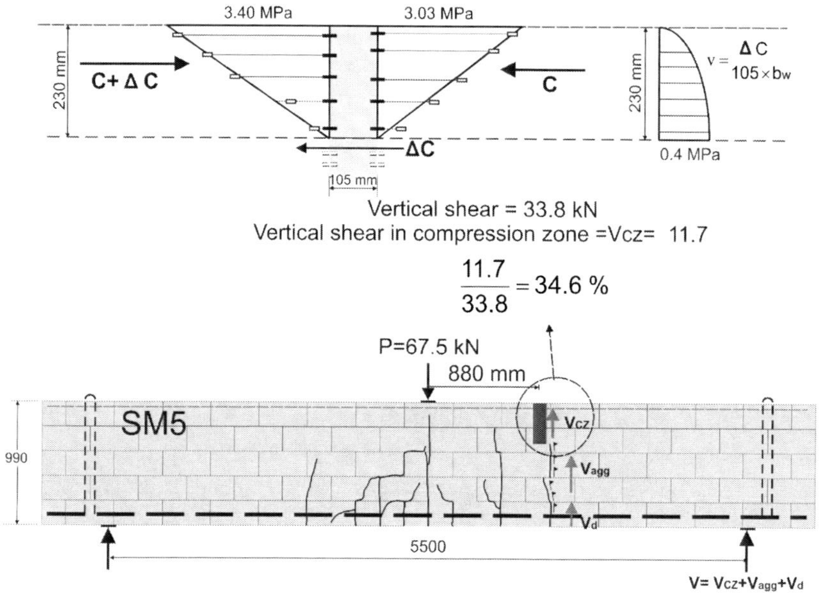

SM5. It can be seen from the figure that the contribution of the compression zone was only 35 %. Although the breakdowns of aggregate interlock in Specimen SM5 was followed by the engagement of strut and tie actions, this mechanism cannot be considered reliable as it did not develop in all the beams. As such, it can be concluded that the major mechanism governing shear strength in RM beams without stirrups is aggregate interlock.

Discussion of Results

THE EFFECT OF REINFORCEMENT STIFFNESS

Table 2 summarizes the major experimental observations of the tested RM beams with North American code predictions. It is worth mentioning that the calculation of experimental shear strengths in **Table 2** was made at a critical section at d from point load. It can be seen in **Fig. 4** and **Table 2** that the use of GFRP rebar (E = 48,200 MPa) in SS5 resulted in a reduction of 40 % in shear strength compared to Specimen SS4 reinforced high-strength steel (E = 200,000 MPa) with a similar reinforcement area. To better understand why the use of low stiffness reinforcement led to such a significant reduction in shear failure strength, consider **Fig. 15a**, which shows the crack patterns in Specimens SS4 and SS5 at an applied load of 38.5 kN (equivalent to a shear stress of 0.34 MPa). It can be noticed that crack spacings in both beams are similar; four major flexural-shear cracks developed through the vertical joints of the bottom course in both beams. However, the crack widths and the level reached by the cracks were significantly different in each beam. In Specimen SS5, the two major cracks close to the point load (one from each side) extended upward and diagonally, passing the mid-depth of the beam. The other two cracks reached to the bed joint and inclined horizontally through the bed joint toward the load point. In Specimen SS4, on the other hand, only one crack reached the mid-depth on the west side close the point load at the same shear stress. The other three cracks barely reached the bed joint. Furthermore, the crack widths in SS5 were considerably greater than those in SS4. This increase in crack widths in SS5 compared to those in SS4 can be explained with the help of **Fig. 15b** and **c**. The application of the load of 38.5 kN caused a mid-span steel strain of 3,608 $\mu\varepsilon$ in SS5 versus only 1,386 $\mu\varepsilon$ in SS4 as shown in **Fig. 15b**. Given that both beams have similar reinforcement ratios, the wider cracks in SS5 appear to be primarily the result of the reduced longitudinal stiffness available to control crack widths. This increase in reinforcement strain in SS5 was associated with increased longitudinal strain at mid-depth of the beam (as shown in **Fig. 15c**), leading to wider cracks. Increasing the widths of the cracks limited their ability to resist shear stresses through aggregate interlock capacity, causing SS5 to fail at a load of 43.12 kN, which is equivalent to 60 % of the failure load of Specimen SS4.

Despite the engagement of arch action and shear key mechanisms, the use of GFRP in SM5 resulted in a reduction of 28 % in shear strength compared to its corresponding Specimen SM4 reinforced with high-strength steel. **Fig. 16** shows that an

TABLE 2 Experimental observations of the tested RM beams with North American code predictions.

	Geometry and Reinforcement					Experimental Observations								Predictions					
														TMS 402		CSA S304		Hoult, Sherwood, Bentz, and Collins	
Beam ID	a/d	d (mm)	b_w (mm)	Steel Type	ρ (%)	f'_m (MPa)	P_{exp} (kN)	V_{exp} [a] (kN)	ν [b] (MPa)	Δ [c] (mm)	Δ/(L/2) [c] (x 10^{-3})	Failure Mode	β [e]	V_{pred} (kN)	V_{exp} / V_{Pred}	V_{pred} (kN)	V_{exp} / V_{Pred}	V_{pred} (kN)	V_{exp} / V_{Pred}
SS4	3.10	307	190	Steel	0.30	18.0	72.4	36.7	0.63	9.27	9.76	Diagonal Tension	0.149	46.3	0.79	26.5	1.39	30.6	1.20
SS5	3.14	303	190	GFRP	0.34	17.5	43.1	22.1	0.38	14.68	15.45	Diagonal Tension	0.091	45.3	0.49	14.5	1.52	18.9	1.17
SM4	3.16	880	190	Steel	0.33	18.5	130.8	68.9	0.41	10.23	3.68	Diagonal Tension	0.095	134.5	0.51	58.9	1.17	65.6	1.05
SM5	3.13	880	190	GFRP	0.30	19.5	93.2	50.1	0.3	74.1 (38) [d]	26.6 (9.6) [d]	Strut and Tie + Shear Key	0.068	85.1	0.59	33.2	1.51	41.7	1.20
														Mean	0.60		1.40		1.16
														STDV	0.12		0.13		0.06
														COV (%)	20		9		5

Note: [a] $V_{exp} = \frac{P_{exp}}{2} + V@d$ from point load (due to self weight); [b] $v = \frac{V_{exp}}{b_w d}$; [c] values at ultimate failure; [d] values at the initial breakdown of aggregate interlock capacity; [e] $\beta = \frac{V_{exp}}{b_w d \sqrt{f'm}}$.

FIG. 15 The effect of reinforcement stiffness on cracking behavior of Specimens SS4 and SS5.

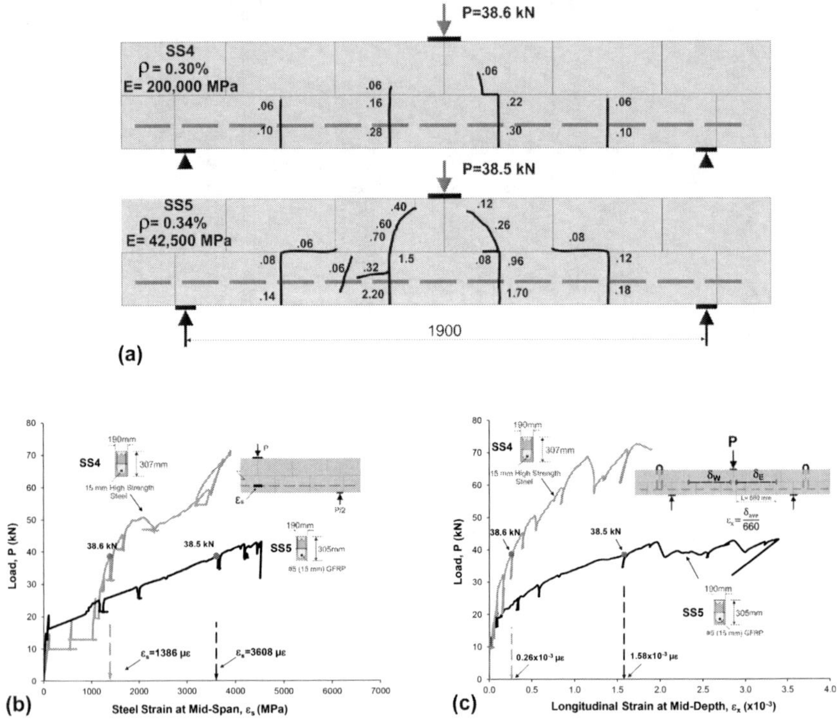

applied load of 60 kN caused only minor cracking in SM4, whereby cracks extended to the second lowest bed joint. In Specimen SS5, however, applying 60 kN led to the formation of a major flexural crack under the point load and three wide flexural-shear cracks. All these cracks extended beyond the mid-depth of the beam, with the crack in the east span inclined to the center of the beam in a jagged manner through the bed and head joints. This can be attributed to the fact that the use of low stiffness reinforcement in SM5 resulted in increasing the mid-depth strain (the longitudinal strain in SM5 was six times that in SM4 at this loading stage as shown in **Fig. 16c**), causing numerous wider cracks. The widening of cracks reduces their ability to resist shear stresses through aggregate interlock. This can explain why the failure of SM5 was initiated by the breakdown of beam action at a lower load than that of SM4.

THE SIZE EFFECT

It can be seen in **Fig. 4** and **Table 2** that increasing the effective depth from 307 mm to 880 mm resulted in a 36 % reduction in the failure shear stress. **Fig. 17** shows the crack patterns in SS4 and SM4 at a shear stress of 0.34 MPa. Note that in this figure

FIG. 16 The effect of reinforcement stiffness on cracking behavior of Specimens SM4 and SM5.

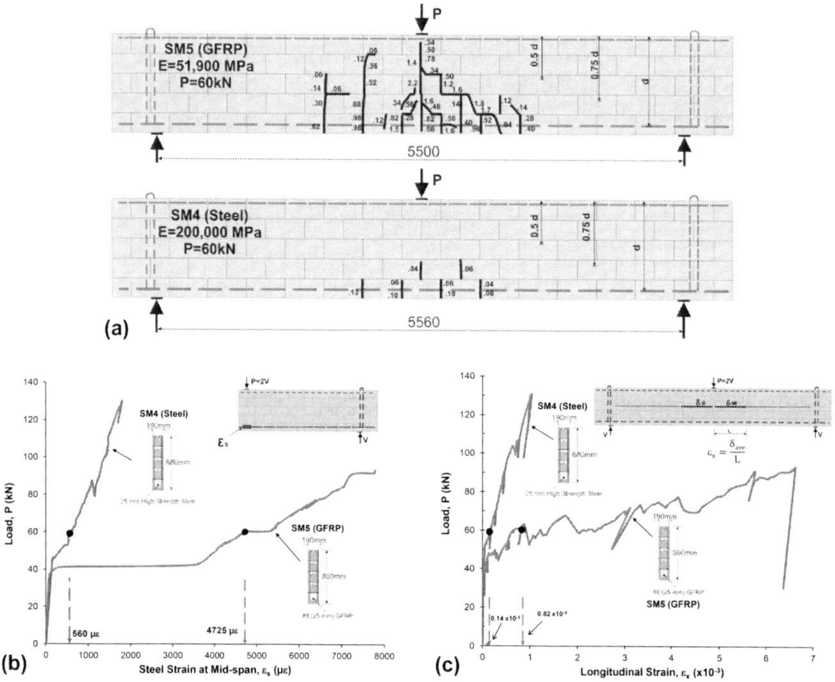

FIG. 17 Crack patterns of Specimens SS4 and SM4 at shear stress of 0.34 MPa.

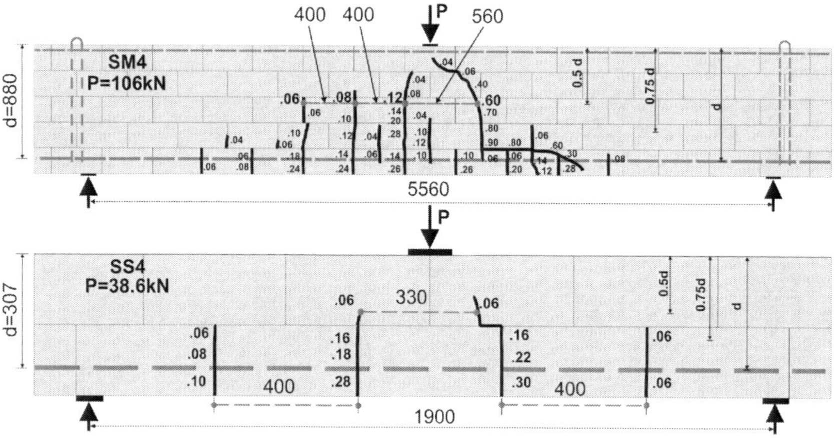

specimens have been drawn such that the effective depths of both beams are the same size on the page. The horizontal crack spacing at mid-height in Specimen SS4 was governed by the distance between the head joints (400 mm in this case) as these joints offer predefined paths for cracks to develop and propagate through. In Specimen SM4, on the other hand, the average crack spacing at mid-depth was 453 mm, which is about half of the effective depth (equivalent to 0.53 d). This trend is similar to that observed in crack spacings at mid-height in reinforced concrete beams [25]. This increase in crack spacings was associated with significant increases in crack widths. **Fig. 17** shows that the maximum crack width at mid-depth of SM4 is eight times that in SS4. Because the capacity of cracks to transfer shear stress decreases as the crack width increases, it appears that the reduced aggregate interlock capacity in the larger specimens was the primary cause of the reduced shear strengths.

The size effect was also observed in GFRP-reinforced beams. Increasing the effective depth from 303 mm in SS5 to 880 mm in SM5 caused a 23 % reduction in the failure shear stress. Analysis of crack patterns of these beams at shear stress 0.21 MPa (**Fig. 18**) showed that the crack spacing at mid-depth increased slightly with the increase in *d* and that the average crack spacings at mid-height of SM5 was 400 mm, which is equivalent to 0.5 d. This slight increase in crack spacing was associated with significantly increased crack widths. The widest crack in SM5 was 3.6 times greater than that in Specimen SS5 at this shear stress level. These wider crack widths then precipitate shear failure at a lower shear stress due to reduced aggregate interlock capacity.

FIG. 18 Crack patterns in SS5 and SM5 at shear stress of 0.21 MPa.

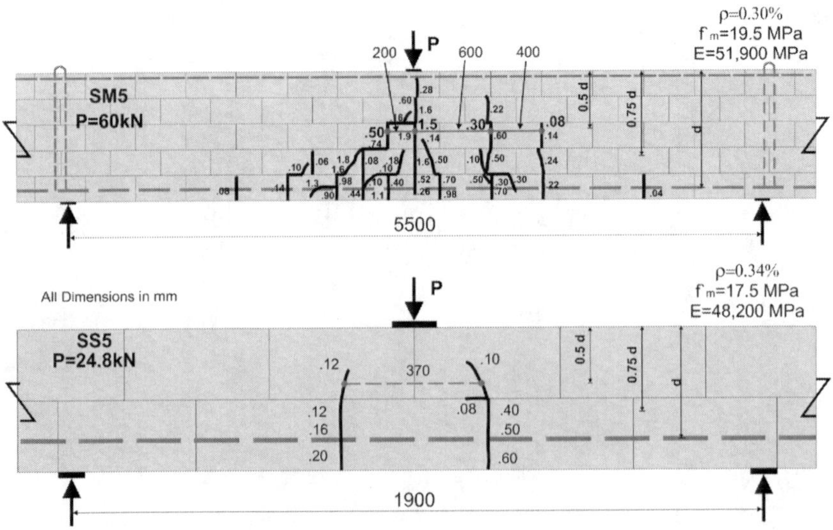

The crack spacing results observed in this paper agree with the results of a major phase of the extensive experimental program conducted at Carleton University [16], which was devoted to studying the size effect in conventional steel RM beams. It was noted that the crack spacing at the mid-depth section increases with the increase in effective depth (size) of the beams and that the crack spacing in large RM beams (five-courses or more) is a function of the effective depth of the beam in a similar fashion to reinforced concrete beams.

CODE PREDICTIONS

The normalized failure shear stresses of Specimens SS4, SS5, SM4, and SM5 are plotted versus the effective depth in Fig. 19 along with the predictions of the codes considered in this paper. It can be seen that the TMS 402 code considerably overestimated the shear strength of these beams. The shear strengths of the GFRP-reinforced specimens, in particular, were poorly predicted. This poor estimation can be attributed to the fact that TMS 402 code does not account for the reinforcement stiffness (E), effective depth, or the reinforcement ratio. The CSA S304 code

FIG. 19 Failure shear stresses for RM beams with different depths.

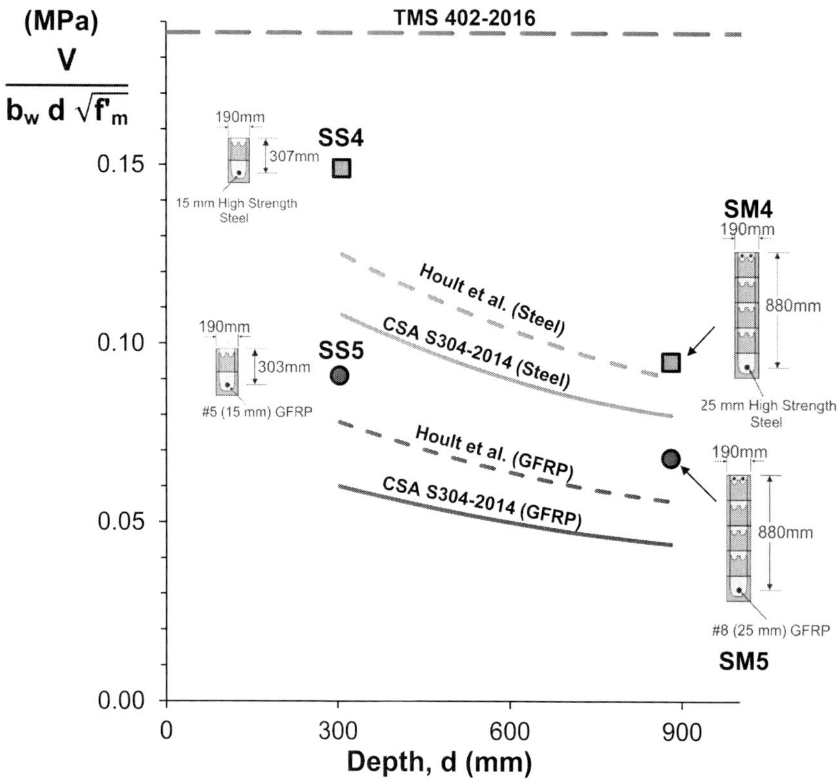

and the Hoult, Sherwood, Bentz, and Collins method, on the other hand, were able to safely capture the variations in the shear strength of RM beams with different reinforcement types and different effective depths. The Hoult, Sherwood, Bentz, and Collins method gave more accurate predictions because it was derived specifically to deal with high strains, which is the case for RM beams reinforced with GFRP rebars. While confirming that the limits on yield strength and material type are entirely appropriate, the results of this section highlight the necessity of revising the TMS 402 code to address the strain effect and the size effect on the shear strength of RM. The general method of the CSA S304 code can serve as a basis for this revision.

The predictions of failure shear stresses in **Table 2** were made based on the masonry compressive strength obtained from testing the associated prisms of each beam. Building and testing masonry prisms typically is not practical due to time and cost constraints, including the lack of local availability of high-capacity testing machines and the complexities of transporting the samples. As such, masonry design codes provide tabulated values of the specified compressive strength (f'_m) based on the compressive strength of the block and the type of mortar. It is also important to note that TMS 402 has a very different method of verifying masonry assembly compressive strength compared to CSA S304. A question can be raised as to how the predictions of the masonry design codes would be if f'_m values were based on tabulated characteristic values. The tabulated characteristic f'_m for Specimen SM4 is 20.7 MPa according to TMS 402 versus 13.5 MPa according to CSA S304. Consider the failure shear stress of Specimen SM4, which failed at a shear force of 68.9 kN. TMS 402 predicts the beam to fail at 134.5 kN based on prism tests (corresponding to a $Vexp/Vpred$ of 0.51) and to fail at 142.7 kN based on tabulated f'_m (making a $Vexp/Vpred$ value of 0.48). On the other hand, The $Vexp/Vpred$ ratios for CSA S304 and the Hoult, Sherwood, Bentz, and Collins method increase from 1.17 and 1.05 based on prism tests to 1.31 and 1.12, respectively, based on tabulated f'_m. Thus, when tabulated f'_m are used in predicting the failure shear stresses, the TMS 402 code gives more non-conservative predictions and CSA S304 becomes more conservative. The method of Hoult, Sherwood, Bentz, and Collins will still be less conservative than CSA S304 when f'_m are based on tabulated values.

The accuracy of Hoult, Sherwood, Bentz, and Collins in predicting the shear strength of GFRP-RM beams suggests the importance of integrating this method into the CSA S304. However, more experimental research is needed to verify the accuracy of the method in predicting the shear strength GFRP-RM beams with different reinforcement ratios.

Concluding Remarks

Based on the results of this experimental investigation, the following conclusions can be drawn:

- The fundamental shear behavior of the RM beams is similar to that of reinforced concrete beams.
- The major mechanism of shear transfer in RM without shear reinforcement is aggregate interlock.

- The beams reinforced with GFRP rebars failed at significantly lower shear stresses than the equivalent beams reinforced with steel.
- The failure shear stresses of RM beams decreased significantly due to increases in depth indicating that size effect is real in RM.
- As shear provisions of TM S402 code are intended to be used for masonry beams reinforced with conventional steel only, they produced predicted failure shear stresses that do not match the experimental behavior of the beams reinforced with GFRP or high-strength steel tested in this investigation. Engineers who want to design RM elements should be fully aware of the limits on yield strength and material type specified by the masonry design codes.
- The general method of the CSA S304-2014 code and the Hoult, Sherwood, Bentz, and Collins method provided safe predictions of the shear strength of GFRP-RM with the Hoult, Sherwood, Bentz, and Collins method giving more accurate predictions.
- The results of this paper highlight the necessity of revising masonry design codes to address the effect of reinforcement stiffness (types) and the size effect. The general method of CSA S304 can serve as a base for such revisions.

References

[1] Galal, K. and Enginsal, M. A., "Flexural Behavior of GFRP-Reinforced Concrete Masonry Beams," *J. Compos. Constr.*, Vol. 15, No. 1, 2011, pp. 21–31.

[2] Brick Industry Association, "Technical Note No.17B: Reinforced Brick Masonry," BIA, Reston, VA, 1999.

[3] Dhanasekar, M. and Wong, K., "Evaluation of Shear Capacity Equation for Masonry Beams without Web Reinforcement," *Proceedings of the Sixth Australasian Masonry Conference*, Adelaide, South Australia, July 12–13, 2001, M. Griffith, Ed., Adelaide University, Adelaide, South Australia, 2001, pp. 115–124.

[4] Kiss, R. M., Kollar, L. P., Jai, J., and Krawinkler, H., "FRP Strengthened Masonry Beams," *J. Compos. Mater.*, Vol. 36, No. 5, 2002, pp. 521–536.

[5] Galal, K. and Sasanian, N., "Out-of-Plane Flexural Performance of GFRP-Reinforced Masonry Walls," *J. Compos. Constr.*, Vol. 14, No. 2, 2010, pp. 62–174.

[6] Tariq, M. and Newhook, J. P., "Shear Testing of FRP Reinforced Concrete without Transverse Reinforcement," *Proceedings of the Annual Conference of the Canadian Society for Civil Engineering*, Moncton, Canada, 2003, Paper No. GCF-340-1-GCF340-10.

[7] Razaqpur, A. G. and Isgor, O. B., "Proposed Shear Design Method for FRP-Reinforced Concrete Members without stirrups," *ACI Struct. J.*, Vol. 103, No.1, 2006, pp. 93–102.

[8] El-Sayed, A. K., El-Salakawy, E., and Benmokrane, B., "Shear Strength of FRP-Reinforced Concrete Beams without Transverse Reinforcement," *ACI Struct. J.*, Vol. 103, No. 2, 2006, pp. 235–243.

[9] Bentz, E. C., Massam, L., and Collins, M. P., "Shear Strength of Large Concrete Members with FRP Reinforcement," *J. Compos. Constr.*, Vol. 14, No. 6, 2010, pp. 637–646.

[10] Khaja, M. N. and Sherwood, E. G, "Does the Shear Strength of Reinforced Concrete Beams and Slabs Depend upon the flexural Reinforcement Ratio or the Reinforcement Strain?" *Can. J. Civil Eng.*, Vol. 40, No. 11, 2013, pp. 1068–1081.

[11] CSA Committee S304, "Design of Masonry Structures, CSA S304-14," Canadian Standards Association, Toronto, ON, 2014.

[12] The Masonry Society, *TMS 402/602-16, Building Code Requirements and Specification for Masonry Structures and Related Commentaries*, The Masonry Society, Longmont, CO, 2016.

[13] Bentz, E. C., Vecchio, F. J., and Collins, M. P., "Simplified Modified Compression Field Theory for Calculating Shear Strength of Reinforced Concrete Elements," *ACI Struct. J.*, Vol. 103, No. 4, 2006, pp. 614–624.

[14] Walraven, J. C., "Fundamental Analysis of Aggregate Interlock," *J. Struct. Div.–ASCE*, Vol. 107, No. 11, 1981, pp. 2245–2270.

[15] Hoult, N. A., Sherwood, E. G., Bentz, E. C., and Collins, M. P, "Does the Use of FRP Reinforcement Change the One-Way Shear Behavior of Reinforced Concrete Slabs?" *J. Compos. Constr.*, Vol. 12, No. 2, pp.125–133.

[16] Sarhat, S. R., "The Size Effect and Strain Effect in Reinforced Masonry," Ph.D. thesis, Carleton University, Ottawa, ON, 2016.

[17] CAN/CSA-A165, *CSA Standards on Concrete Masonry Units*, Canadian Standards Association, Mississauga, ON, 2004.

[18] CSA A179-04, *Mortar and Grout for Unit Masonry*, Canadian Standards Association, Mississauga, ON, 2014.

[19] ASTM C109/C109M-16a, *Standard Test Method for Compressive Strength of Hydraulic Cement Mortars (Using 2-in. or [50-mm] Cube Specimens)*, ASTM International, West Conshohocken, PA, 2016, www.astm.org

[20] ASTM C1019-18, *Standard Test Method for Sampling and Testing Grout*, ASTM International, West Conshohocken, PA, 2018, www.astm.org

[21] ASTM C1552-16, *Standard Practice for Capping Concrete Masonry Units, Related Units and Masonry Prisms for Compression Testing*, ASTM International, West Conshohocken, PA, 2016, www.astm.org

[22] Fenwick, R. C. and Paulay, T., "Mechanisms of Shear Resistance of Concrete Beams," *J. Struct. Div.–ASCE*, Vol. 94, No. 10, 1968, pp. 2235–2350.

[23] Taylor, H. P. J., *Investigation of the Forces Carried across Cracks in Reinforced Concrete Beams in Shear by Interlock of Aggregates*, TRA 42.447, Cement and Concrete Association, London, 1970.

[24] Kani, M. W., Huggins, M. W., and Wittkopp, R. R., *Kani on Shear in Reinforced Concrete*, University of Toronto Press, Toronto, ON, 1979.

[25] Sherwood, E. G., "One-Way Shear Behaviour of Large, Lightly Reinforced Concrete Beams and Slabs," Ph.D. thesis, University of Toronto, ON, 2008.

[26] Jourawski, D. J., "Sur le Résistance d'un Corps Prismatique et d'une Piece Composée en Bois ou on Tôle de Fer à une Force Perpendiculaire à leur Longeur [Remarks on the Resistance of a Prismatic Body and a Composite Part Made of Wood or Sheet Iron to a Force Perpendicular to their Length]," *Annales des Ponts et Chausees*, Memoires and Documents, 3rd Series, Vol. 12, Part 2, 1858, pp. 328–351.

[27] Kani, G. N. J., "How Safe Are Our Large Reinforced Concrete Beams?" *J. Am. Concrete I.*, Vol. 64, No. 3, 1967, pp. 128–141.

[28] Gergely, P., "Splitting Cracks along the Main Reinforcement in Concrete Members," Report to Bureau of Public Roads, U.S. Department of Transportation, Cornell University, Ithaca, NY, 1969.

STP 1612, 2018 / available online at www.astm.org / doi: 10.1520/STP161220180012

Charles B. Clark, Jr.,[1] Cortney L. Fried,[1] and James Bryja[2]

History of Cold-Formed Steel Framing Used as a Backing for Brick Veneer

Citation

Clark, C. B., Jr., Fried, C. L., and Bryja, J., "History of Cold-Formed Steel Framing Used as a Backing for Brick Veneer," *Masonry 2018, ASTM STP1612*, N. V. Krogstad and W. M. McGinley, Eds., ASTM International, West Conshohocken, PA, 2018, pp. 158–180, http://dx.doi.org/10.1520/STP161220180012[3]

ABSTRACT

The last 50 years have seen many changes in the design and installation of the typical anchored brick veneer/cold-formed steel framing wall assembly. During that time, many groups have published and revised their design guidelines and recommendations for such assemblies. An extensive amount of testing and research has been undertaken to better understand different aspects of the assembly. This paper outlines the history of anchored brick veneer/cold-formed steel framing by specifically reviewing the testing and research conducted, the technical papers developed, and the changes in industry guidelines. Changes in wind load provisions during this period are also reviewed. This information serves as the background for the paper, "Addressing Maximum Design Deflection for Cold-Formed Steel Framing When Used as a Backing for Brick Veneer," which presents a revised approach for determining maximum deflection for cold-formed steel framing supporting brick veneer.

Keywords

brick veneer, cold-formed steel, deflection, cracking, wind load, ties, anchors, steel stud

---

Manuscript received February 16, 2018; accepted for publication April 2, 2018.

[1]Brick Industry Association, 12007 Sunrise Valley Dr., Suite 430, Reston, VA 20191, USA C. B. C. http://orcid.org/0000-0003-2158-2073, C. L. F. http://orcid.org/0000-0001-8510-4666

[2]General Shale Products, 3015 Bristol Hwy., Johnson City, TN 37601, USA http://orcid.org/0000-0003-2138-4210

[3]ASTM International Symposium on *Masonry 2018* on June 26, 2018 in San Diego, CA, USA.

Introduction

Prior to the late 1960s, when brick veneer was installed on a commercial building or a large multifamily residential building, it was positioned over a backing of masonry or concrete. In 1968, United States Gypsum (USG) released an alternate backing wall system that used the company's gypsum sheathing in combination with cold-formed steel framing.

Using the cold-formed steel framing system permitted early enclosure of the building during construction and offered benefits related to the reduced weight of the system compared to a similar masonry backing wall. As a result, the system became a popular choice. However, little was known about the nature and behavior of the brick veneer/cold-formed steel wall system at the time, and engineers expressed concern that the stiff veneer with a flexible backing would result in poor performance.

Unfortunately, in the early years of their implementation, unsatisfactory performance was observed in brick veneer/cold-formed steel walls with reports of severe cracking, water leakage, spalling brick, and failed anchors/ties. Engineers began research into the wall system to better understand its behavior and to provide more guidance regarding design and detailing requirements.

This paper presents the evolution of understanding the system behavior and the design recommendations for the system in a chronological format, with parameters relevant to each period of time grouped together.

Period 1: Late 1960s to 1980

USG is considered the developer of the wall system; however, many other cold-formed steel framing manufacturers developed their own versions of the system [1]. Recommendations were initially provided for deflection limits of L/240 and L/360, in accordance with typical wall deflection criteria as presented in building codes. Detailing recommendations were also provided in the manufacturer's literature.

PERIOD 1 INDUSTRY RECOMMENDATIONS

The 1968 USG manual included the following regarding the system:

- The use of this system was limited to non-load-bearing applications only.
- Twenty-gauge metal studs were used for stucco and masonry exteriors.
- Gypsum wallboard or lath/plaster were used for interior finishes.
- The use of the system was limited to wind loads up to 15 psf. Limiting heights and stud spacing dimensions were provided in tables.
- The figures in the document depicted the use of corrugated anchors/ties with spacing of anchors 24 in. vertically and 16 in. horizontally.

Notable changes to the USG manual during this period included the following:

- In 1969, the limiting heights were reduced, and the system was identified for use with wind loads up to 40 psf. Composite action of the studs with the gypsum sheathing was considered in determination of stiffness and deflection.
- In 1973, the wall system was reported to withstand 90 mph wind pressure per testing.

- In 1976, a recommendation for applying a 15-lb felt building paper over the gypsum sheathing was added to account for the potential of moisture penetration in "high wind areas." The recommended anchors were 22-gauge corrugated anchors/ties.
- In 1977, the number of stud sizes and gauges for the system was expanded and the deflection of L/360 included the brick veneer stiffness. Anchor spacing was reduced to 16 in. on center. The use of Portland cement-lime mortar was recommended. Language requiring a qualified architect or engineer to evaluate product use was added.
- In 1980, the maximum wall height was reduced from the 1979 values. The deflection limitation was changed to L/600 for the entire wall assembly.

In Technical Note 28B dated January 1979 [2], the Brick Institute of America (now the Brick Industry Association [BIA]) identified and explained the system issues or challenges that designers should consider, as well as providing some recommendations and suggested details. These recommendations included:

- Using a maximum allowable cold-formed steel stud deflection of L/600 to L/720, with L/900 as the preferred deflection criteria.
- Providing securely attached sheathing on both sides of the metal studs.
- Providing doubled studs at the jambs of all windows, doors, and other openings in the system and horizontal bracing of the studs at mid-height.
- Prohibiting the use of corrugated anchors/ties. Minimum nine-gauge corrosion-resistant adjustable wire anchors/ties were recommended instead, with a requirement to anchor through the sheathing to the studs, not into the sheathing only. Spacing of these wire anchors was limited to 24 in. maximum horizontally and vertically with one anchor/tie per 2 2/3 ft^2.
- Requiring the use of Portland cement-lime mortars for all buildings over three stories and recommending the use of Type S mortar where wind loads would exceed 25 psf.
- Using severe weathering grade brick (Grade SW) per ASTM C62, *Standard Specification for Building Brick (Solid Masonry Units Made From Clay or Shale)* [3], ASTM C216, *Standard Specification for Facing Brick (Solid Masonry Units Made from Clay or Shale)* [4], or ASTM C652, *Standard Specification for Hollow Brick (Hollow Masonry Units Made From Clay or Shale)* [5]. Used and salvaged brick were prohibited.

When this technical note was revised the following year [6], the BIA removed the preference for an allowable cold-formed steel stud deflection of L/900, leaving the recommendation of L/600 to L/720 and noting that "[t]he Metal Lath/Steel Framing Association [ML/SFA], which represents the manufacturers of metal studs, does not agree with this recommendation. Research is planned to study this question" [6].

PERIOD 1 TESTING AND TECHNICAL PAPERS

With respect to the brick veneer/cold-formed steel (BV/CFS) system, no significant testing occurred, and no notable technical papers were published during this period.

PERIOD 1 WIND LOADS

Prior to 1972, requirements for wind loads were distributed among regional building codes as well as codes developed by individual cities. Depending on the area in the United States, a site could be under the provisions of the International Conference of Building Officials Uniform Building Code, the Building Officials and Code Administrators International National Building Code, or the Southern Standard Building Code. Many of these codes identify wind pressure as a single value based on map location and building height during this period. For instance, the 1967 Uniform Building Code identifies an allowable resultant wind pressure from a map of the United States based on a 30-ft height above ground. This value is modified by tabulated values for heights less than 30 ft or greater than 49 ft in order to obtain a design value. Wind load design values are provided for various height zones. No difference between positive and negative pressure is noted; therefore, the value is assumed to apply to wind loads in both directions. Loads are intended to be applied to the vertical projection of the wall within each zone.

The first consensus standard for structural loadings was issued by the American National Standards Institute (ANSI) in 1972. ANSI Standard A58.1 was titled, "Building Code Requirements for Minimum Design Loads in Buildings and Other Structures." In this standard, the wind speed contours and values presented reflect a probabilistic 50-year average recurrence period based on fastest-mile wind speeds by the U.S. National Weather Service (collected since the 1930s).

PERIOD 1 DISCUSSION

Significant changes in the USG literature occurred every year, including critical parameters such as maximum wall heights and allowable wind load pressures. This suggests the changes may have been a reaction to situations occurring in the field. The manufacturers assumed that stud and gypsum sheathing will act compositely to resist load and deflection. The manufacturers' literature also referenced the "non-composite addition of the brick veneer stiffness" in determining deflection of the wall system with limits of L/360, which would permit the use of shallow, more flexible stud sections.

USG's inclusion of a deflection criteria of L/600 in 1980 appears to be a compromise between the recommendations of the BIA and ML/SFA. The BIA recommended L/600 for the stud deflection while the similar USG requirement applied to the entire wall assembly, resulting in a stud deflection closer to ML/SFA recommendations. The two groups (the BIA and ML/SFA) agreed to pursue research to help decide the issue. Certain recommended parameters from the BIA technical notes were never adopted into the manufacturer literature, such as using wire-type anchors/ties only. Not until 1976 do manufacturers recommend a water-resistive barrier but only for certain situations. Given that the gypsum sheathing was relied upon to achieve composite action with the studs, moisture ingress should have been a concern in all conditions because wet or damp gypsum sheathing would be detrimental to the intended strength and performance of the system.

Though removed from all later versions of the technical note, the BIA's deflection criterion preference in the 1979 version is notable for its correlation to other known serviceability criteria. Based on a simplified analysis, a deflection of L/900 with a nominal 4-in. veneer corresponds with a crack width of approximately 0.015 in., which is the upper bound of allowable cracking in concrete [7]. The L/720 criteria results in an estimated crack width of 0.02 in. using the same method.

Period 2: 1981 through 1985

During this period, industry literature continues to be modified, with some testing performed internally by manufacturers. The first significant testing would also be performed in 1981. This research project was performed at Clemson University, cosponsored by the BIA and ML/SFA, in an attempt to reach consensus regarding deflection recommendations. This is the testing program alluded to in the 1980 update of BIA Technical Note 28B.

PERIOD 2 INDUSTRY RECOMMENDATIONS

In 1983, USG modified the published details to include adjustable anchors/ties and building felt and required the first anchor/tie to be installed a maximum of 8 in. from discontinuities. Use of masonry cement mortars was prohibited. In 1985, a revised stud selection guide table was published, with limiting heights based on USG's internal testing, BIA recommendations, and the results of the Clemson study.

PERIOD 2 TESTING AND TECHNICAL PAPERS

- 1981 USG Testing [8]: Four wall assemblies using 4-in. nominal brick veneer were tested as part of this internal program. The walls were rated for an allowable load of 11 psf, using 3.5-in. deep, 22-gauge, 11.25-ft long studs. Each of the walls was subjected to 100 cycles of loading at 15 psf of both positive and negative pressure. After completion of the cyclic loading, each panel was tested to failure, with failure defined as excessive deflection under minimal load increase. It was reported that all four assemblies resisted loads in excess of 45 psf, with one reaching 60 psf. The maximum panel deflection at the time of failure was approximately 1.5 in.
- 1981 National Concrete Masonry Association (NCMA) Testing [9]: NCMA tested four 8-ft tall wall assemblies during this period, two with 4-in. nominal concrete masonry veneer and two with 4-in. nominal brick masonry veneer. The cold-formed steel assembly was rated for 40 psf per the manufacturer's literature. Panels were tested with a single application of load to failure, which was defined as loss of tensile bond. Failure occurred at an average pressure of 31 psf, less than the rated pressure of the cold-formed steel assembly. The panels were loaded beyond this failure point until the anchors/ties buckled at 80 psf and the studs buckled at approximately 100 psf. USG took exception to the results of this testing [10], though the NCMA data plots indicate a change in slope at approximately 30 psf, which was consistent with the failure definition.
- 1981/1982 Clemson Research [11]: Arumala and Brown conducted a two-phase experiment on the performance of BV/CFS wall assemblies. The first

phase tested two anchor/tie types: corrugated sheet metal anchors/ties and "DW-10 drywall adjustable" anchors/ties. The DW-10 anchors/ties consist of a wire bent in a triangular shape inserted through a vertical loop in a backing plate. Based on the results of the first phase of testing, the DW-10 anchors/ties were selected for use in the second phase. In the second phase, six full-size wall panels (each 4 ft wide by 9.33 ft tall) were constructed and subjected to pressure (three samples positive, three samples negative) as well as water permeance testing per ASTM E514, *Standard Test Method for Water Penetration and Leakage Through Masonry* [12]. To supplement the experimental testing, mathematical models were developed to permit evaluation of different support conditions, such as anchor/tie stiffnesses, relative wall stiffnesses, and the inelastic behavior of the wall system.

- The wall assemblies included concrete spandrel elements, which were part of the frame developed to support the test specimens. This resulted in shorter stud heights compared to the height of brick (7 ft 10 in. versus 9 ft. 4 in.).
- The load testing and water testing was performed in an alternating fashion, with water permeance testing occurring after the wall was loaded. Prior to any load testing, the as-built samples were subjected to an initial water permeance test as a baseline.
- Test loads consisted of design pressure (24.16 psf), two times the design pressure (48.32 psf), and three times the design pressure (72.48 psf).
- Measured deflections in both the brick and backing were less than L/360 under design load. Under two times the design load, only one wall had brick deflections that exceeded L/360, and under three times the design load, one wall had deflections close to L/360, with the remaining exceeding L/360.
- One of the positively loaded walls cracked between 26 psf and 31 psf. Two of the negatively loaded walls did not crack at three times the design load. The remaining three walls cracked between 52 psf and 62 psf. The compressible filler at the top of the wall permitted translation of the veneer at this location, particularly under negative pressures, which likely reduced stress on the veneer.
- The two walls that cracked at loads lower than those of the other walls absorbed a higher percentage of water during the permeance testing. Moisture was observed at the gypsum sheathing during the testing. There were difficulties in maintaining the chamber seal during the water testing, so the results are not definitive; however, the water penetrating the veneer was not correlated with the magnitude of applied load.
- The mathematical models accurately predicted the failure loads. The cold-formed steel stud assembly appeared to reduce the load carried by the veneer by 44 %. Any composite action that developed between the sheathing and the studs did not significantly alter the load carried by the brick veneer.
- The adjustable wire anchor/tie was recommended for use in the system in lieu of corrugated anchors/ties. A version with a heavier gauge backing was recommended for larger design loads. The forces in the anchors/ties were found to be nonuniform; the current methods of load distribution severely underestimate the maximum anchor/tie forces obtained. No anchors/ties failed during testing, but the backing plates underwent significant deformation at the top line of anchors/ties under negative pressure loading.

- It was determined that the use of flexural rigidity does not accurately allocate the proportion of load to each wythe.
- With respect to the current system of relying on the deflection of the cold-formed steel stud alone, it was recommended that evaluation of the support conditions and the flexural stiffness of the brick veneer be considered during design.

In 1985, Bell and Gumpertz [13] reported that perceptions of special interest influence assemblies that did not adequately represent the full range of systems installed to date, and failure to rationally describe the behavior of the system prevented the acceptance of the findings. They performed a simple idealized analysis of the system but recommended performing a finite element analysis in the future. Other recommendations included: disregarding contributions from the sheathing when evaluating the backing wall, prohibiting corrugated metal anchors/ties, designing the connection and anchorage of the backing wall framing elements (the stud to track connections were identified as a concern), and sealing anchor/tie penetrations to prevent moisture ingress. Based on the analyses performed and the research to date, maintaining stud deflection criteria of L/600 to L/720 was recommended with the understanding that it would not necessarily result in acceptable performance in all situations.

Bell and Gumpertz also recommended further testing so that a rational model of behavior under wind loading could be developed using reinforced concrete (load resistance factor design) as a guide. The proposed research topics included:

- Behavior of the system after cracking
- The effect of veneer cracks on water penetration, anchor/tie engagement, and masonry deterioration under cyclic loads
- Brick anchor/tie strength and stiffness
- Contributions of the gypsum sheathing
- Collection of data regarding performance of in-service wall assemblies.

PERIOD 2 WIND LOADS

The concepts of main wind force resisting systems and components/cladding were introduced to the ANSI Standard A58.1 in 1982. It incorporated the understanding of "…the aerodynamics of wind pressure around building corners, eaves and ridge areas, as well as the effects of area averaging on pressures" [14]. In 1985, the American Society of Civil Engineers (ASCE) assumed control of the ANSI A58.1 standard, which would now be identified as ASCE 7 [15].

PERIOD 2 DISCUSSION

As indicated by Bell and Gumpertz, the results of testing during this period were not widely accepted. Despite a goal of reaching consensus on a recommendation for deflection limitations, both the brick and cold-formed steel framing industries maintained their previous stances, with the ML/SFA releasing a publication recommending a maximum stud deflection of L/360 [16] and the BIA recommending a maximum stud deflection of L/600 to L/720. The Clemson testing indicated that the behavior of the wall system is more complex than originally expected, and a rational basis could not be extrapolated from the experimental results. Therefore, the approach of limiting stud deflection to minimize cracking was maintained.

Although not providing definitive answers, the Clemson testing identified several characteristics of the wall system that either led to updated design recommendations or future testing:

- Distribution of load between wythes according to flexural stiffness did not reflect actual conditions.
- Composite action between the sheathing and studs did not contribute substantially.
- Nonuniform load distribution on anchors/ties resulted in significantly higher forces observed at the topmost line of anchors/ties.
- Corrugated anchor/tie stiffness is inadequate and inappropriate for this system.
- Water permeance was not correlated with load pressure or presence of cracking, although moisture was noted at the gypsum sheathing at some locations.
- Boundary conditions play a role in the performance of the system.

Period 3: 1986 through 1989

In 1986, the Canada Mortgage and Housing Corporation (CMHC) initiated a large-scale research and education plan to address outstanding structural and building science concerns with BV/CFS wall systems. A nationwide survey was first conducted to identify how the system was being designed and installed for use in identifying problems and research topics. Other testing during this period generally proceeded with the purpose of addressing the research topics suggested by Bell and Gumpertz.

PERIOD 3 INDUSTRY RECOMMENDATIONS

- USG literature was not significantly changed during this period.
- An updated version of BIA Technical Note 28B was released in February 1987 [17]. It maintains the recommendation for deflection limit of L/600 and recommends a minimum 18 ga stud thickness and galvanizing of both studs and anchors/ties.
- BIA Technical Note 44B on anchors/ties was released in March 1987 [18]. It recommends limits for mechanical play and deformation in anchors/ties. Mechanical play in anchors/ties should not exceed 0.05 in., and all anchors/ties should not deform more than 0.05 in. under 100 lb load in either tension or compression. Variations in strength and stiffness when anchors are at the extended range of adjustability were also discussed. A revised version of this technical note was released in September 1988, incorporating minor revisions and edits including anchor/tie spacing and air space dimensions.
- The Canadian masonry code CAN3-S304-M84 permits the veneer to be designed rationally; this is justified by the increased portion of lateral load the

veneer carries prior to initiation of cracking compared to other configurations of brick veneer. Allowable tensile stresses normal to the mortar bed joint are 0.25 MPa (36.25 psi) [19].

PERIOD 3 TESTING AND TECHNICAL PAPERS

- 1986 Results of the CMHC Survey [20]: This report presented the results of the survey used to identify how the system was being used and designed. It was sent to architects, structural engineers, masonry contractors, cold-formed steel/drywall contractors, building officials, and building inspectors. More than 50 % used L/360 as deflection criteria for the wall system; 20 % used L/600. Architects were responsible for selecting/designing the system in the majority of cases, with the structural engineer of record designing the system only 30 % of the time. Shop drawings were rarely required. The available information about the system was noted to be inadequate by 80 % of the respondents.
- 1986 Testing at the University of Alberta [21]: This testing focused on load-deflection behavior of both components of BV/CFS wall systems and the overall system.
  - A variety of anchor/tie types were tested independently, including multiple gauges of corrugated sheet metal anchors/ties and multiple shapes of wire anchors/ties.
  - The backing walls were tested independently as well. The specimens consisted of 18-gauge studs with 0.5-in. gypsum sheathing fastened to both sides of the studs. Additional specimens consisting of deeper 20-gauge studs were also tested to compare with the 18-gauge walls.
  - Sixteen full-size assemblies (brick veneer and cold-formed steel combined) were tested under positive lateral pressure. The anchor/tie spacing patterns, cavity width, anchor/tie type, and stud type were varied over each specimen. During testing of these walls, deflections were measured at both the brick veneer and the backing wall.
  - Buckling of the lighter gauge corrugated anchors/ties (22- and 24-gauge) occurred at significantly lower loads than other anchors/ties tested. The heavier gauge corrugated anchor/tie (16-gauge) did not buckle but caused crushing in the gypsum sheathing. The T-shaped and ladder anchors/ties also crushed the gypsum sheathing.
  - Uneven load distribution to the anchors/ties was confirmed, along with variations in performance based on the connection between the anchor/tie and the cold-formed steel stud.
  - It was found that wall deflection was reduced when more anchors/ties were located at the top of the wall. In addition, all combinations of walls and anchors/ties exhibited reserve strength after the veneer cracked, except those using the 22- and 24-gauge corrugated anchors/ties.
- 1988 Influence of Tie Flexibility, Relative Length, and End-Boundary Condition on Brick Veneer-Metal Stud Flexural Bond Stress [22]: The intent of this paper was to demonstrate how a BV/CFS system can perform well in light of the reported failures by analyzing variations in anchors/ties, stud length, and stud end fixity. It determined that in some cases as-built conditions could have

permitted better performance in existing walls than would be suggested by analysis. Fixity at the base connection achieved L/1445 in a stud sized for L/600 deflection. Achieving fixity at the connection was noted to require specific design effort.

- 1988 Masonry Veneer Wall Systems—Structural Engineering Report 156 [23]: This report documented a University of Alberta program that tested anchors/ties attached to full height wall assemblies and full height wall assemblies under positive and negative lateral load to establish recommended performance (phi) factors. Testing confirmed that anchors/ties must be designed to transfer loads exceeding those based on tributary area and that the connection between the anchor/tie and the stud could affect the stiffness and strength of the anchor/tie. Analytical models (two- and three-dimensional) were developed. The test walls demonstrated significant load carrying capacity post-cracking. Cracks developed in multiple bed joints, but the location of horizontal cracks generally did not align with anchor/tie locations and anchors/ties remained engaged.
- 1989 Computech Engineering Services Analysis of Lateral Loads on Anchored Brick Veneer on Metal Studs [24]: This paper performed finite element analyses of the BV/CFS. Two stages of analysis were performed. The first stage analyzed buildings with various structural systems that included components representing the cladding. The second stage was a more detailed analysis of the veneer, anchors/ties, and studs including both full-height panels and spandrels. The analysis indicated that designers should expect the veneer to crack under service loads. The veneer generally cracked at two locations around mid-height, four courses apart. Once formed, no further cracks occurred. Maximum load applied was approximately two times the cracking load. Crack size was approximately 0.02 in. when two cracks occurred. A lower bound deflection of L/360 was considered acceptable.

Secondary forces occurred in anchors/ties after cracking. Stiffer studs increased the load at which first cracking occurs. Based on the analyses performed, the design criteria for the metal stud design is rational, but the prediction of anchor/tie forces is neither rational nor conservative. The current code provision does not account for the uneven distribution of forces in the anchors/ties. A simplified procedure for determining maximum anchor/tie force is included.

PERIOD 3 WIND LOADS

No significant revisions occurred to the wind load provisions of ASCE 7 during this period. The 1988 version of ASCE 7 was released, but it included only minor edits from the previous version in 1982 when it was still under the ANSI nameplate [14].

PERIOD 3 DISCUSSION

The results of the 1986 CMHC survey are notable. Most of the designers used a stud deflection of L/360. Only 20 % used L/600 for deflection, even though recommendations for the more stringent deflection existed (but may not have been widespread) more than five years earlier. This may be related to the fact that architects were the primary designers/selectors of the systems, and they were highly reliant on manufacturers' published literature, which may or may not have included L/600 as an option in the tables or other design aids. There was wide consensus (80 %) on the need for more information about the system and direction on how to design it.

During this period, both experimental and computer-based studies were performed to confirm findings of previous work and expand knowledge of the system behavior. The Computech study performed by Kelly, Goodson, Ronald, and Asher [24] analyzed wall behavior, most notably for walls taller than those used in previous studies. The simulations in this study suggested that two parallel cracks would occur at mid-height in most cases, with single cracks at the others. Kelly, Goodson, Ronald, and Asher [24] predicted veneer cracks with openings up to 0.040 in. if a single crack forms at mid-height but does not opine about the acceptability of a crack that size, deferring to requirements for reinforced concrete and recommending a design deflection of L/360. The smaller crack openings resulting when two cracks form (0.02 in.) were slightly larger but consistent with the upper bound of acceptable concrete cracking. It was indicated that the cracks will close after removal of load; however, since the cracks are noted to occur under service loads, it would suggest that the cracks will be subjected to cyclic opening and closing, with potential accelerated deterioration as a result.

Testing and analysis during this period provided significant insight into the anchor/tie behavior and justified the prohibition of corrugated anchors/ties. Multiple anchor/tie types were tested or analyzed. Initial experimental testing evaluated the anchors/ties alone, but later, anchors/ties were tested in combination with studs as the open cross-section of the stud contributed to the anchor/tie behavior.

It was found that the anchors/ties at the top of the wall section carry approximately 40 % of the total lateral load prior to cracking after which the load is shared with the anchor/tie(s) closest to the crack that formed. This variation in loading is not accounted for in the accepted methodology of assigning load to anchors/ties based on tributary area.

A procedure for determining the anticipated upper bound of anchor/tie forces was proposed in the Computech study. The procedure calculates the largest anchor/tie force for both cracked and uncracked simple span veneer conditions. The largest value of these two cases is compared to the tributary lateral load of two times the brick veneer weight (per Chapter 14 of the IBC), with the largest value of all three cases governing for anchor/tie selection or design. This procedure would be later incorporated into a design guide for brick veneer/cold-formed framing systems.

Testing related to the water penetration resistance of system was not pursued during this period.

Period 4: 1990 through 2002

The multiyear research program for both structural and building science aspects of the system resulting from the CMHC survey was conducted at McMaster University [16]. The recommendations from the testing program results were eventually presented in publications released by CMHC starting in 1990.

PERIOD 4 INDUSTRY RECOMMENDATIONS

- As a result of the McMaster University testing on the BV/CFS system, Canada changed requirements in revisions to the 1995 version of the National Building Code. The National Building Code references the following standards: S304.1-94, *Masonry Design for Buildings, Limit States Design*; A370-94, *Connectors for Masonry*; and CAN3-A371-94, *Masonry Construction for Buildings*. The code includes two methods for evaluating the wall deflection: (1) Using a limit of L/600 but including the mechanical play and deformation of the anchors/ties in addition to the deflection of the stud itself (A370 provides limits for the extent of allowable mechanical play and deformation in the anchors/ties); or (2) using a limit of L/720 for stud design but with stricter allowable mechanical play and deformation limits. Anchors/ties were required to be designed for 40 % of the tributary lateral load on the stud.
- The Western States Clay Products Association (WSCPA) released "Design Guide for Anchored Brick Veneer over Steel Studs" in January 1995. It recommended criteria of L/360 for cold-formed steel deflections instead of the BIA-recommended L/600. The WSCPA noted that the L/360 deflection results in crack widths between 0.02 in. and 0.04 in. However, the WSCPA also acknowledged that building performance can be categorized into two levels: "Level 1 for institutional buildings with a long design life…[and] Level 2 for commercial buildings with an average design life" [25]. This implies that more stringent deflection requirements are likely appropriate for an institutional building intended for Level 1 performance. A procedure for determining maximum anchor/tie forces per Kelly, Goodson, Ronald, and Asher [24] was included.
- The BIA released a revised Technical Note 28B in November 1999 [26]. The organization of the document was restructured in this version, with less focus on the past poor performance of the system and more focus on the detailing of the current system. The deflection limit of L/600 was maintained and noted to minimize crack widths.

PERIOD 4 TESTING AND TECHNICAL PAPERS

- "Fabrication and Testing of Components of Steel Stud Backup Walls" [19] is Part 1 of the comprehensive research program conducted at McMaster University. The experimental testing included stud-to-track connections and full-size wall assemblies with and without sheathing. Analysis, including evaluation of experimental results, was also performed. A minimum stud thickness of 18-gauge was recommended. Cracking of the brick veneer was considered inevitable, but stud deflections of L/720 were recommended to limit crack widths. When a crack width limit of 0.2 mm (0.008 in.) was applied, the resulting deflection criteria was L/900. Gypsum sheathing provided no more than a 17 % increase in stiffness due to composite action, which was quickly lost during repeated load cycles. However, it did provide sufficient bracing to permit the studs to develop full moment capacity when applied to both stud flanges. Use of bridging was recommended. Anchors/ties installed coincident with a web cutout in the stud resulted in failure caused by local buckling.

- "Water Permeability of Cracked Masonry Veneer" (Drysdale/Kluge/Roscoe) is Part 2 of the comprehensive research program conducted at McMaster University. According to the CMHC Best Practice Guide [27], small-scale masonry samples were tested to determine potential correlation between cracking and water leakage. The results were inconclusive; however, a decrease in leakage rate was noted over the duration of the test along with some partial autogenous healing of cracks.
- "Performance of Brick Veneer/Steel Stud Wall Systems Subject to Temperature, Air Pressure, and Vapour Pressure Differentials" [28] is Part 3 of the comprehensive research program conducted at McMaster University. Five partial height wall assemblies were constructed and subjected to simulated environmental conditions. The testing determined that air barriers are critical to the performance of the assembly. Small breaches can permit a significant amount of air, which carries water vapor. However, perfect air barrier installation was not achievable, even in the lab.
- "Behaviour of Brick Veneer/Steel Stud Tie Systems" [29] is Part 4 of the comprehensive research program conducted at McMaster University. A detailed evaluation of wall anchors/ties was performed, with twelve different anchor/tie styles evaluated under full cooperation from the anchor/tie manufacturers. BIA recommendations for mechanical play limits and deformation under a given load were considered and determined to be appropriate. Testing confirmed that anchors/ties must be designed to transfer loads exceeding those based on tributary area. In addition, the stiffness and strength of the anchor/tie could be affected by the connection between the anchor/tie and the cold-formed steel stud. Connections to stud flanges can cause deformations under heavy loads based on the comparison with tests of web-mounted anchors/ties.
- "Tests of Full Scale Brick Veneer Steel Stud Walls to Determine Strength and Rain Penetration Characteristics" [30] is Part 5 of the comprehensive research program conducted at McMaster University. Five full-scale wall systems were tested within a custom test apparatus to permit both structural loading and water testing of the same specimens. A wall specimen (Wall 1) represented the typical BV/CFS assembly, with support at the top and bottom of the wall and no side restraint. Water testing and load testing were performed on the assembly with both a pressurized and unpressurized air space/cavity. All walls consisted of 18-gauge studs, 8.5 ft tall. The stud size selected for all the walls was based on analysis that indicated a deflection of L/360 and L/720 under loads of 42 psf and 21 psf, respectively. The initial cracking in Wall 1 occurred at approximately 29 psf. This corresponded with a horizontal crack forming near mid-height. Displacement at the top of the wall increased after cracking, but the panel was still capable of resisting additional loading, which was attributed to the incidental restraint provided by the expansion joint at the top of the wall. A second horizontal crack occurred at 79 psf. The rate of deflection increased after this point; however, failure of the wall did not occur until 150 psf, when the cold-formed steel studs failed near the midpoint due to local flange buckling, which was attributed to the large anchor/tie forces at this location. Evidence of flexural-torsional buckling was not observed. Based on the

findings, a stud deflection of L/720 was recommended to limit the average crack size to those similar to reinforced concrete as well as consideration of track flexibility and anchor/tie parameters.

PERIOD 4 WIND LOADS

In 1995, ASCE 7 changed the basis of the wind speed maps from fastest mile to 3-s gust due to the elimination of collecting fastest mile wind speeds by the U.S. National Weather Service. A review of the archived data found that gust speed had been recorded for a significant number of weather stations along with fastest mile wind speed, which would allow ASCE to maintain comparison to future wind speeds with previous years.

Using ASCE 7 during this period resulted in an increase in the basic wind speed compared to previous codes. Other changes included the addition of a coefficient to account for topography, a reduction in the internal pressure coefficient, and the consolidation of the edge and corner zones. However, in the commentary of ASCE 7-95 for Appendix B—Serviceability Considerations, the load combination for use in checking lateral deflections included a factor of 0.7 on the design wind load, which is indicated to correspond with a wind event that has a 5 % annual probability of being exceeded [31]. The legacy building codes did not include detailed information regarding deflection requirements.

In 2000, the first edition of the International Building Code was released. It referenced ASCE 7-98 for the wind load requirements, except for the simplified design method. The IBC provides direction regarding serviceability requirements of wall systems in Section 1604.3—Serviceability. Table 1604 OF ASCE 7 lists the deflection limits for various building elements. Footnote (f) in the deflection limit table indicates that the components and cladding wind load are permitted to be multiplied by 0.7 in order to evaluate deflections [32].

PERIOD 4 DISCUSSION

At the beginning of this period, the large-scale testing at McMaster University was still ongoing. It included experimental testing not only on the structural aspects of the wall system but also on water penetration resistance and building science issues (air movement, water vapor, condensation). By the end of this period, the findings of the large-scale testing program were finalized and incorporated into the Canadian building codes as well as into guides for use by designers and specifiers in Canada and the United States.

The National Building Code of Canada changed its previous cold-formed steel stud deflection recommendation of L/360 to L/720 based on the results of the McMaster testing. In the United States, there are no code requirements, but disagreement over stud deflection still existed during this period. The BIA maintained their deflection recommendation of L/600. The WSCPA's design guide identified a deflection of L/360 but acknowledged that some types of construction, such as institutional buildings, require more stringent criteria. An increase of minimum cold-formed steel stud gauge to 18-gauge (43 mils) from 20-gauge (33 mils) was recommended for long-term durability.

Some composite action between the gypsum sheathing and the cold-formed steel studs was found to occur but is highly variable based on installation. In addition, cyclic loading rapidly decreased the extent of composite behavior that existed, indicating that it should not be considered for flexural capacity or increased stiffness in the selection or evaluation of a cold-formed steel stud section for the wall assembly. However, gypsum sheathing was found to provide adequate bracing for the cold-formed steel stud flanges assuming the spacing of fasteners occurred within the limits tested (6 in. on center), but use of intermediate bridging is also recommended. Bridging connections are critical to long-term performance of the backing wall and are recommended to be mechanically fastened to the studs using clip angles. Rigid insulation serving as sheathing provided some bracing for the cold-formed steel studs but to a lesser degree than gypsum board sheathing and did not permit development of full moment capacity of the cold-formed steel studs [19].

When full-size wall samples were tested, significant reserve capacity was observed. After cracking initially, the system resisted loads up to 80 psf before a second parallel crack formed. Even at failure (local buckling of the cold-formed steel studs), where the load applied was 150 psf, the integrity of the connection of the veneer to the backing wall was maintained.

Beyond the variability of loads applied to anchors/ties, the anchor/tie configuration and components play a significant role in the behavior of the veneer. In order to provide adjustability, performance can be compromised. At the extremes of the adjustable range, there can be loss of stiffness in the anchor/tie due to deformation of the components. Adjustability also involves the combination of a minimum of two components; therefore, gaps between components exist and movement of the veneer will occur before the components contact each other and engage. This "take-up" in the adjustable anchors/ties contributes to the flexibility of the veneer—but not the stud—and needs to be accounted for when analyzing the system.

The highly loaded anchors/ties in the system impart concentrated loads to the cold-formed steel studs that are not currently accounted for because the cold-formed steel studs are assumed to carry a uniformly distributed load. These concentrated loads can lead to local buckling effects in the cold-formed steel studs. Failures frequently occurred when anchor/tie attachment points corresponded with prefabricated web cutouts in the cold-formed steel stud.

Water penetration testing performed during this period provided somewhat inconclusive results. The McMaster University testing subjected full-size wall specimens to combined water testing and pressure [30]. Water penetration was observed but should not have been unexpected because it was understood at the time that a single wythe of brick masonry was not watertight. Based on the test results, creation of a pressure-equalized rainscreen was recommended to minimize the potential for water penetration beyond the drainage cavity.

The building science testing performed as part of the McMaster University program identified that the placement of components and their integrity was critical to good performance [28]. Condensation occurred in a number of specimens tested, which suggests that some of the corrosion and material degradation issues observed in the publicized failures could have been related to condensation as well as to failure of the drainage cavity to adequately direct bulk water out of the system.

During this period, the code documents explicitly permitted using reduced loads for deflection calculations. The IBC model code allowed the application of a reduction factor to its deflection limit charts. A similar reduction was also noted and discussed in the non-mandatory Commentary Appendix C of ASCE 7.

Period 5: 2003 through the Present

PERIOD 5 INDUSTRY RECOMMENDATIONS

- In 2003, the American Iron and Steel Institute (AISI) released its first design guide on this topic, CF03-1 [33]. As part of this document, it included the following serviceability limits:
    - Stud flexural deflection limit = L/600.
    - Maximum total free mechanical play for anchors/ties = 0.050 in.
    - Maximum deflection for anchors/ties = 0.050 in. under 100 lb of load (tension or compression).

Recommendations for anchors/ties included connecting directly to the stud to avoid compression of the sheathing under load and connecting to the stud web instead of the flange to avoid axially loaded screws.

The anticipated maximum crack width is the combination of the stud deflection, half the mechanical play of the anchors/ties, and the anchor/tie deformation under load. Generally, this value is divided by two to obtain an average crack width because the cross section of the crack is triangularly shaped, ranging from 0 in. wide at the compression face to the maximum crack width at the tension face.

- BIA Technical Note 44B regarding anchors/ties was updated in 2003 [34].
- The WSCPA released a second edition of their "Design Guide for Anchored Brick Veneer over Steel Stud Systems" in August 2004 [35]. It maintained its recommended criteria of L/360 for cold-formed steel deflections instead of the BIA-recommended L/600. One significant difference in this edition is the recommendation of anchor/tie stiffness between 2 kips/in. and 6 kips/in.
- Canada updated their code design standard for masonry in 2004, releasing the Canada Standards Association's CSA S304.1-04, in which the required backing wall deflection was reduced from L/600 or L/720 to L/360. These recommendations were incorporated into the 2005 version of the Canadian National Building Code.
- The BIA released a revised Technical Note 28B for brick veneer/steel stud walls in December 2005, which is the current version [36]. Primarily, it reorganized previous content and added a summary of recommendations, in addition to incorporating minor edits. The recommendation for cold-formed steel deflection of L/600 remains.

- The Canadian Sheet Steel Building Institute incorporated the Canadian National Building Code deflection limit change into their literature, releasing a technical bulletin [37] prior to revising their "Steel Stud Brick Veneer Design Guide."
- The AISI released a revised version of the design guide, "Brick Veneer Cold-Formed Steel Framing Design Guide, D112-13," in 2013 [16].
  - The provisions were similar to the previous version (CF03-1) in maintaining a deflection recommendation of L/600.
  - Includes maximum stud spacing of 16 in. on center, with variations allowed per design or anchor/tie spacing.
  - Indicates anchors/ties should be designed to carry up to 40 % of the load tributary to a given stud.

PERIOD 5 TESTING AND TECHNICAL PAPERS

- In 2005, "New CSA S304.1-04 Design of Masonry Structures" was presented at the Tenth Canadian Masonry Symposium [38]. This paper summarized the numerous changes in the code document, which included:
  - Reduced allowable deflection of backing construction of L/360, with the caveat that, "Greater stiffness of the structural backing may be necessary depending on the design of the wall's moisture management system to handle rainwater that enters the cavity" [38].
  - Minimum flexural bond strength of 29 psi for mortar.
  - Required a row of top anchors/ties, with one attached to each stud. Elsewhere, anchors/ties are permitted to be staggered. The previous supplemental restrictions on mechanical play and anchor/tie deformation to use a less stringent deflection were also removed.
  - No justification was provided for the changes.
- In 2006, the Canadian Sheet Steel Building Institute issued a technical bulletin regarding the new National Building Code of Canada, which implemented the code changes [37]. In this document, justification for the deflection reduction was provided. It indicates that other methods of mitigating potential water ingress can be as successful or more successful than controlling crack width, such as:
  - Pressure moderation of the air space.
  - Minimized mortar droppings in air space.
  - Fully filled and tooled mortar joints.
  - Adequately spaced veneer expansion joints with appropriate joint sealant.
  - Durable flashings with appropriately detailed and constructed seams, end dams, drips, and functional weeps.
  - Design elements that provide protection, such as cornices or overhangs.
- In 2007, Arumala summarized the methodology to date regarding the use of the system in the United States, which has not significantly changed [39].
  - Indicated a stud deflection of L/360 is a widely accepted standard for low-rise buildings, while L/600 is reserved for taller structures.
  - He noted that the system is not inherently flawed, just highly susceptible to problems due to moisture ingress. Good design will equal good performance.

PERIOD 5 WIND LOADS

- A major change to the wind load provisions occurred in ASCE 7-10. To provide consistency with the seismic provisions, the basic wind speed maps were modified to larger values that were appropriate for direct use in limit states design/load resistance factor design. The load combinations were changed such that the load factors for wind are 1.0, and if service level loads are desired for allowable stress design or serviceability checks, the calculated values would be multiplied by 0.6.
- Three basic wind speed maps were provided in lieu of the previous single basic wind speed map. Each of the new maps incorporated the risk level/importance factor of the building in question and was associated with a mean recurrence interval (MRI) for a wind event. Lower risk structures used a wind event with a lower MRI to determine the required design wind pressure.
- For serviceability purposes, the Commentary Appendix C of ASCE 7-10 provided alternate basic wind speed maps for lower MRI values. It noted that using the design maps from the main body of the code is "excessively conservative" [40]. Selection of the appropriate MRI map is left to the engineer; however, the commentary suggests an MRI of 10 years for a typical building and an MRI of 50 to 100 years for buildings that may be sensitive to drift.
- The National Building Code of Canada recommends the use of 1/10 year wind loads for design of cladding [41].
- During this period, the IBC maintained the deflection limits in Table 1604.3. In IBC versions that reference ASCE 10 and later, the reduction factor in footnote (f) includes the multiplier to convert the strength design formula output to an unfactored load, changing the factor of 0.7 to 0.42. The load reduction in the IBC for deflection calculations is not intended to be applied concurrently with the load reduction in the commentary of ASCE 7 [42].

PERIOD 5 DISCUSSION

Early in this period, AISI, the overarching organization representing cold-formed steel framing in the United States, released their first design guide for BV/CFS wall systems. The design guide included a recommendation for stud deflections of L/600 and the provisions from BIA documentation recommending restrictions on mechanical play and deformation of anchors. The design guide also recommended web-mounted anchors/ties, which is consistent with the findings of experimental testing in the previous period, where high anchor forces could initiate pullout of screws attached to the stud flange and loaded axially. The web-mounted anchors/ties permit the use of additional fasteners and result in a loading configuration where sheet metal screws have an increased capacity.

Until 2005, the United States and Canada generally agreed on the deflection recommendations. At that time, Canada revised their National Building Code and reduced the deflection requirements to L/360. Per the paper presented at the Tenth Canadian Masonry Symposium, these changes were among nearly 30 changes to the masonry provisions in the code. Explanations were not provided but were later clarified in a CSBBI technical bulletin that stated that there were more effective methods of limiting moisture ingress (not unique to BV/CFS applications) than restricting deflection. This suggests that the primary concern with the BV/CFS system was not structural in nature or related to cracking as a direct serviceability concern. The assertion that the primary concern is moisture ingress, not structural concerns, is echoed by Arumala in his 2007 paper [39].

It is interesting to note that the modification to the Canadian code prohibits the use of masonry cement mortars as well as type N mortars on ungrouted masonry due to the new minimum flexural bond strength of 29 psi.

The change to the wind loads in ASCE 7-10 was significant, but when the strength design formula output loads are reduced to service level loads by the new load combination factors, they are similar in magnitude to previous code versions. The conversion to strength design loads resulted in design MRI events that were too conservative for serviceability loads (700 years for routine structures). The design loads in earlier codes were based on 50-year MRI events.

Summary

Brick veneer on cold-formed steel framing began as an industry-developed innovation with the engineering community scrambling to understand and explain the behavior of the system as many changes and adaptations were occurring. Stringent deflection recommendations were an understandably conservative response to the lack of available data and the inconsistent performance observed in the field.

Since that time, experimental testing and computer modeling have demonstrated that the system behavior is deceptively complex. For instance, the load distribution between the veneer and backing is not uniform and depends on the stiffness of the anchors/ties as much as on the relative stiffnesses of the brick and the backing. The distribution of load to the anchors/ties can be nonuniform despite assumptions otherwise. In addition, the system has significant reserve capacity, with local buckling of the cold-formed steel occurring prior to loss of engagement with the veneer. This knowledge, gained over the years through research and implemented into modern designs, has led to improved and consistent performance of brick veneer/cold-formed steel framing.

References

[1] Ramirez, I., "Structural Performance of Brick Veneer/Steel Stud Wall Systems as Designed in Current Practice," M.S. thesis, University of Texas at Austin, Austin, TX, 1988.

[2] Brick Institute of America, "Brick Veneer Panel and Curtain Walls–28B," *Technical Notes on Brick Construction,* Brick Institute of America, Reston, VA, 1979.

[3] ASTM C62-17, *Standard Specification for Building Brick (Solid Masonry Units Made From Clay or Shale)*, ASTM International, West Conshohocken, PA, 2017, www.astm.org

[4] ASTM C216-17a, *Standard Specification for Facing Brick (Solid Masonry Units Made from Clay or Shale)*, ASTM International, West Conshohocken, PA, 2017, www.astm.org

[5] ASTM C652-17a, *Standard Specification for Hollow Brick (Hollow Masonry Units Made From Clay or Shale)*, ASTM International, West Conshohocken, PA, 2017, www.astm.org

[6] Brick Institute of America, "Brick Veneer Panel and Curtain Walls–28B," *Technical Notes on Brick Construction,* Brick Institute of America, Reston, VA, 1980.

[7] American Concrete Institute, "ACI 224R—Control of Cracking in Concrete Structures," American Concrete Institute, Farmington Hills, MI, 2001.

[8] Marchello, M. J., "U.S.G. Tests Provide New Data for Determining Limiting Heights of Steel-Stud/Brick-Veneer Curtain Wall Assemblies," *Form & Function,* Issue 3, USG Corporation, Chicago, IL, 1981, pp. 14–15.

[9] National Concrete Masonry Association, "Laboratory Investigation of Metal Stud Back Up Systems for Masonry Veneers," National Concrete Masonry Association, Herndon, VA, 1981.

[10] Marchello, M. J., "Steel Stud Curtain Wall Design Stands Up to Rigorous Tests," *Form & Function,* USG Corporation, Chicago, IL, 1984.

[11] Arumala, J. O. and Brown, R. H., "Final Report—Performance Evaluation of Brick Veneer with Steel Stud Backup," Department of Civil Engineering, Clemson University, Clemson, SC, 1982.

[12] ASTM E514, *Standard Test Method for Water Penetration and Leakage Through Masonry,* ASTM International, West Conshohocken, PA, 2014, www.astm.org

[13] Bell, G. R. and Gumpertz, W.H., "Engineering Evaluation of Brick Veneer/Steel Stud Walls, Part II—Structural Design, Structural Behavior, and Durability," *Proceedings of the Third North American Masonry Conference,* Arlington, TX, June 3–5, 1985, J. H. Matthys and J. G. Borchelt, Eds., University of Texas at Arlington, Arlington, TX, 1985, pp. 5-1–5-20.

[14] Ghosh, S. K., "Searching for Simplicity: The Evolution of Wind Provisions in Standards and Codes in the United States," *RCI Interface*, Vol. 26, No. 3, 2008, pp. 13–21.

[15] Mehta, K. C., "Wind Load History: ANSI A58.1-1972 to ASCE 7-05," *Proceedings of the 2010 Structures Congress,* Orlando, FL, May 12–15, 2010, S. Senapathi, K. Casey, and M. Holt, Eds., ASCE, Reston, VA, 2010, pp. 2134–2140.

[16] American Iron and Steel Institute, "Brick Veneer Cold-Formed Framing Design Guide—D112," American Iron and Steel Institute, Washington, DC, 2013.

[17] Brick Institute of America, "Brick Veneer Steel Stud Panel Walls—28B," *Technical Notes on Brick Construction,* Brick Institute of America, McLean, VA, 1987.

[18] Brick Institute of America, "Wall Ties for Brick Masonry—44B," *Technical Notes on Brick Construction,* Brick Institute of America, McLean, VA, 1987.

[19] Drysdale, R. and Breton, N., "Part 1—Strength and Stiffness Characteristics of Steel Stud Backup Walls Designed to Support Brick Veneer," Department of Civil Engineering and Engineering Mechanics, McMaster University, Hamilton, ON, 1991.

[20] Suter Keller, Inc., "Brick Veneer/Steel Stud Wall Design and Construction Practice in Canada—Results of a 1986 Survey," Canada Mortgage and Housing Corporation, Ottawa, ON, 1986.

[21] McGinley, W. M., Longworth, J., Warwaruk, J., and Hatzinikolas, M., "Masonry Veneer and Steel Stud Curtain Walls," *Proceedings of the 4th Canadian Masonry Symposium*, Fredericton, New Brunswick, June 2–4, 1986, J. L. Dawe, R. M. Francis, and A. J. Valsangkar, Eds., University of New Brunswick, Department of Civil Engineering, Fredericton, New Brunswick, 1986, pp. 730–743.

[22] Chin, I. R., Krogstad, N. V., and Monk, C. B., "Influence of Tie Flexibility, Relative Length, and End-Boundary Condition on Brick Veneer-Metal Stud Flexural Bond Stress," *Masonry: Materials, Design, Construction, and Maintenance, ASTM STP992*, H. A. Harris, Ed., ASTM International, West Conshohocken, PA, 1988, pp. 96–117, http://dx.doi.org/10.1520/STP27263S

[23] McGinley, W., Warwaruk, W., Longworth, J., and Hatzinikolas, M., "Stuctural Engineering Report 156—Masonry Wall Veneer Systems," Department of Civil Engineering, University of Alberta, Edmonton, Alberta, 1988.

[24] Kelly, T., Goodson, M., Ronald, M., and Asher, J., "Analysis of the Behavior of Anchored Brick Veneer on Metal Stud Systems Subjected to Wind and Earthquake Forces," *Proceedings of the Fifth North American Masonry Conference*, University of Illinois at Urbana-Champaign, June 3–6, 1990, D. P. Abrams, Ed., Masonry Society, Boulder, CO, 1990, pp. 1359–1370.

[25] KPFF Consulting Engineers, "Design Guide for Anchored Brick Veneer Over Steel Studs," Western States Clay Products Association, Orinda, CA, 1995.

[26] Brick Industry Association, "Brick Veneer: Brick Veneer/Steel Stud Walls—28B," *Technical Notes on Brick Construction,* Brick Industry Association, Reston, VA, 1999.

[27] Canada Mortgage and Housing Corporation, "Best Practice Guide—Building Technology—Brick Veneer Steel Stud," Ottawa, ON, 1996.

[28] Drysdale, R. G. and Kluge, A., "Part 3—Performance of Brick Veneer/Steel Stud Wall Systems Subject to Temperature, Air Pressure, and Vapour Pressure Differentials," Canada Mortgage and Housing Corporation, Ottawa, ON, 1990.

[29] Drysdale, R. and Wilson, M., "Part 4—Behaviour of Brick Veneer/Steel Stud Tie Systems," Canada Mortgage and Housing Corporation, Ottawa, ON, 1989.

[30] Drysdale, R. and Wilson, M., "Part 5—Tests of Full Scale Brick Veneer Steel Stud Walls to Determine Strength and Rain Penetration Characteristics," Canada Mortgage and Housing Corporation, Ottawa, ON, 1990.

[31] American Society of Civil Engineers, "ASCE 7—Minimum Design Loads for Buildings and Other Structures," American Society of Civil Engineers, New York, 1995.

[32] International Code Council, "International Building Code," International Code Council, Country Club Hills, IL, 2000.

[33] American Iron and Steel Institute, "Steel Stud Brick Veneer Design Guide—CF03-1," American Iron and Steel Institute, Washington, DC, 2003.

[34] Brick Institute of America, "Wall Ties for Brick Masonry—44B," *Technical Notes on Brick Construction,* Brick Institute of America, McLean, VA, 2003.

[35] KPFF Consulting Engineers, "Design Guide for Anchored Brick Veneer Over Steel Studs," Western States Clay Products Association, Orinda, CA, 2004.

[36] Brick Industry Association, "Brick Veneer/Steel Stud Walls—28B," *Technical Notes on Brick Construction,* Brick Industry Association, Reston, VA, 2005.

[37] Canadian Sheet Steel Building Institute, "Steel Studs as Back-up for Brick Veneer Changes in NBC 2005," *Lightweight Steel Framing Technical Bulletin,* Canadian Sheet Steel Building Institute, Cambridge, ON, 2006.

[38] Laird, D. A., Drysdale, R. G., Stubbs, D. W., and Sturgeon, G. R., "The New CSA S304.1-04—Design of Masonry Structures," *Proceedings of the Tenth Canadian Masonry Symposium*, Banff, Alberta, June 8–12, 2005, S. Lissel, Ed., University of Calgary, Department of Civil Engineering, Calgary, ON, 2005, pp. 312–321.

[39] Arumala, J. O., "Brick Veneer Steel Stud Wall Systems: State-of-the-Art," *TMS Journal,* Vol. 25, No. 1, 2007, pp. 9–19.

[40] American Society of Civil Engineers, "ASCE 7—Minimum Design Loads for Buildings and Other Structures," American Society of Civil Engineers, Reston, VA, 2010.

[41] Griffis, L. G., "Serviceability Limit States Under Wind Load," *Engineering Journal,* American Institute of Steel Construction, Vol. 30, 1993, pp. 1–16.

[42] Swensson, K. D., "Guide to the Design of Building Systems for Serviceability in Accordance with the 2012 IBC and ASCE/SEI 7-10," International Code Council, Country Club Hills, IL, 2013.

Works Consulted

Borough of Camden, "Camden Planning Guide—Basements and Lightwells CPG4," Borough of Camden, London, 2015.

BRE Group, "Assessing Cracks in Houses," http://web.archive.org/web/20180208211908/https://www.bre.co.uk/page.jsp?id=3356 (accessed August 28, 2017).

Brown, R. H. and Arumala, J. O., "An Experimental Study of Brick Veneer with Metal Stud Backup," *Proceedings of the Sixth International Brick/Block Masonry Conference*, Rome, Italy, May 16–19, 1982, ANDIL, Rome, pp. 1411–1427.

Building Research Establishment, "BRE Digest 251—Assessment of Damage in Low-Rise Buildings," Building Research Establishment, Garston, Watford, UK, 1995.

Canadian Sheet Steel Building Institute, "Lightweight Steel Framing Architectural Design Guide," Canadian Sheet Steel Building Institute, Cambridge, ON, 2002.

Drysdale, R. G. and Suter, G. T., "Exterior Wall Construction in High-Rise Buildings—Brick Veneer on Concrete Masonry or Steel Stud Wall Systems," Canada Mortgage and Housing Corporation, Ottawa, ON, 1991.

Gumpertz, W. H. and Bell, G. R., "Engineering Evaluation of Brick Veneer/Steel Stud Walls, Part I—Flashing and Waterproofing," *Proceedings of the Third North American Masonry Conference*, Arlington, TX, June 3–5, 1985, J. H. Matthys and J. G. Borchelt, Eds., The Masonry Society, Longmont, CO, 1985, pp. 4-1–4-16.

International Code Council, "International Building Code," International Code Council, Country Club Hills, IL, 2015.

The Masonry Society, "TMS 402—Building Code Requirements for Masonry Structures," The Masonry Society, Longmont, CO, 2016.

Metal Lath/Steel Framing Association and Brick Institute of America, "Joint Statement on the Clemson University Brick Veneer-Steel Stud Testing Program," Brick Institute of America, McLean, VA, 1982.

Postma, M. A. and Burnett, E. F. P., "Renovation Strategies for Brick Veneer Steel Stud Wall Construction—Task 2," University of Waterloo for Canada Mortgage and Housing Corporation, Waterloo, ON, 1993.

Structural Clay Products Research Foundation, "Research Report Number 9—Compressive, Transverse, and Raking Strength Tests of Four-Inch Brick Walls," Geneva, IL, 1965.

West, M., Fisher, J., and Griffis, L., "AISC Steel Design Guide 3—Serviceability Design Considerations for Steel Buildings," American Institute of Steel Construction, Washington, DC, 2003.

STP 1612, 2018 / available online at www.astm.org / doi: 10.1520/STP161220170176

Charles B. Clark, Jr.,[1] Cortney L. Fried,[1] and James Bryja[2]

Addressing Maximum Design Deflection for Cold-Formed Steel Framing When Used as a Backing for Brick Veneer

Citation

Clark, C. B., Jr., Fried, C. L., and Bryja, J., "Addressing Maximum Design Deflection for Cold-Formed Steel Framing When Used as a Backing for Brick Veneer," *Masonry 2018, ASTM STP1612*, N. V. Krogstad and W. M. McGinley, Eds., ASTM International, West Conshohocken, PA, 2018, pp. 181–197, http://dx.doi.org/10.1520/STP161220170176[3]

ABSTRACT

Through the years, different values have been recommended as the maximum deflection for the design of cold-formed steel framing when used as a backing for brick veneer. The basis for each of these recommendations varied considerably and included criteria such as limiting the extent of cracking in the veneer, limiting the width of cracks in the veneer, limiting moisture intrusion, and "observations, experience, and judgment." The methods by which this information was determined also varied and included analysis, full-scale testing, and field surveys. In addition, the building code-mandated wind load provisions that were frequently used to calculate maximum design deflection have changed over time. This paper reviews and reassesses the different maximum deflection recommendations by examining their basis and other factors. A comparison of the different maximum deflection recommendations is made considering the impact of the changing wind load provisions. Based on this information, a revised approach for determining maximum deflection for cold-formed steel framing used as a backing for brick veneer is recommended.

Keywords

brick veneer, cold-formed steel, deflection, cracking, wind load, ties, anchors, steel stud

---

Manuscript received September 26, 2017; accepted for publication February 8, 2018.

[1]Brick Industry Association, 12007 Sunrise Valley Dr., Suite 430, Reston, VA 20191 C. B. C. http://orcid.org/0000-0003-2158-2073, C. L. F. http://orcid.org/0000-0001-8510-4666

[2]General Shale Products, 3015 Bristol Hwy., Johnson City, TN 37601 http://orcid.org/0000-0003-2138-4210

[3]ASTM International Symposium on *Masonry 2018* on June 26, 2018 in San Diego, CA, USA.

Introduction

For more than half a century, cold-formed steel framing has been used as a backing for anchored brick veneer in the United States and Canada. Over time, this wall system has evolved into a viable construction system that provides successful performance in a wide variety of commercial, industrial, and institutional structures. Prior to the introduction of this system, brick veneer on mid-rise or high-rise structures had been applied to a backing of concrete masonry.

However, there has been lack of consensus in the design and construction industry regarding the amount of acceptable deflection in the brick veneer/cold-formed steel (BV/CFS) wall system. Over the years, various organizations have proposed different deflection recommendations that were developed from the results of analysis, testing, and surveys. This paper reviews significant aspects that form the basis for these deflection recommendations such as anchor/tie behavior, the consequences of veneer cracking, appropriate wind load values for serviceability, and redundancy within the system. These aspects are considered in the development of updated design deflection recommendations that are proposed.

System Description

In an anchored masonry veneer wall, the brick transfers out-of-plane load directly to the backing construction and is not considered to add load-resisting capacity to the wall system. The BV/CFS wall system consists of a nominal 3-in. or 4-in.-wide brick wythe mechanically attached using corrosion-resistant metal anchors/ties to create a prescribed air space between the veneer and the backing construction. The backing consists of cold-formed steel studs spaced at either 16 in. or 24 in. on center, with gypsum sheathing on the exterior face and gypsum board on the interior. The ends of studs are restrained by cold-formed steel tracks fastened to the structural slabs.

Historical Basis

When the BV/CFS system came into use, the body of knowledge pertaining to this specific system was minimal. Design guides at the time provided limited information and were only made available by certain cold-formed steel stud manufacturers [1]. The performance of the system was presumed to be equivalent to other brick veneer walls, but dedicated research to substantiate such performance was lacking. Other veneer wall systems, such as brick with concrete masonry (CMU) backing, had a history of good performance, and their behavior was better understood. As new building stock was constructed with the BV/CFS system and the existing BV/CFS buildings aged, it became apparent that the performance of the BV/CFS system was inadequate [2].

STIFFNESS OF ELEMENTS

The difference in stiffness between the brick veneer and the cold-formed steel backing was identified as a concern, given that the veneer was the stiffer element.

The analysis applied at the time assumed that load would be distributed proportionally between the veneer and the backing by flexural stiffness and the deflections between the elements would be compatible. As a result, it was thought that limiting the deflection of the stud through simple analysis would in turn result in a similar deflection of the brick veneer.

CRACKING

Historically, limiting the deflection of BV/CFS walls was related to preventing veneer cracking, which was a logical extension of the industry's understanding of unreinforced brick masonry behavior. For unreinforced brick masonry, cracking was considered a failure mode because the performance of the wall is dependent on maintaining the flexural capacity of the masonry that would be significantly reduced if cracking occurred near the critical section [3].

Once masonry veneer became prevalent, a similar philosophy was adopted, although cracking was thought to be less of a structural concern and more of a serviceability concern [4]. The primary structural purpose of veneer construction is to support its self-weight and transfer lateral load to the underlying structure [5]. The grid of anchors/ties used to connect the brick veneer to the backing results in a structurally indeterminate element, which remains stable when cracked, eliminating the structural concerns typically associated with unreinforced masonry.

Although masonry veneer is a drained wall system, the air space or drainage cavity behind the veneer is not intended to receive large quantities of water. Eliminating or reducing veneer cracking will limit the amount of moisture that reaches the drainage cavity and potentially cross to the backing where it could possibly lead to leakage into the building interior or corrosion of the studs and anchors/ties. Today, the use of improved detailing, enhanced anchor/tie design and installation, and more robust water-resistive barrier materials within the wall assembly have reduced the risk of increased moisture reaching the backing. Consequently, cracks larger than those traditionally deemed acceptable may now be considered nonproblematic from a moisture ingress perspective.

DESIGN GUIDANCE CHANGES

As a result of inadequate performance of the BV/CFS system in the field, significant changes to design guidance provided by cold-formed steel manufacturers occurred on a yearly basis. These changes included increases in the maximum allowable wind loads, increases in the minimum gauge of corrugated ties, reducing anchor/tie spacing, and recommending 15-lb felt building paper moisture barriers in the interior of the cavity [1]. The steel stud manufacturers originally recommended stud deflections of L/360, and these were maintained during this time.

Eleven years after the BV/CFS system was introduced to the market, the Brick Industry Association (BIA) published design guidance for this system that recommended using adjustable wire ties instead of corrugated ties, a reduction in anchor/tie spacing, and a more stringent stud backing deflection criterion (L/600). The BIA also indicated that more research was needed to determine proper design criteria and to understand the water penetration resistance of the system [6].

The various industry recommendations for deflection are summarized in **Table 1**.

Behavior

EARLY RESEARCH

Early testing [7] showed surprising results, with the wall assemblies withstanding lateral loads up to three times the design pressure of stud walls designed for a deflection of L/360. This demonstrates a significant reserve capacity of the system. The L/360 deflection was not exceeded at the design load for all the walls tested and only exceeded this limit on two of the six wall specimens after loading exceeded twice the design load. These results obviously did not match the predicted values. Many organizations did not accept the results of the testing and maintained their existing stance [4]. Where wall configurations did not match the test specimen construction, the decision to use L/360 as a deflection limit was left to the professional judgment of the designer [8].

The water penetration testing performed as part of these tests did not show correlation between water ingress and lateral load. However, the results could be considered inconclusive due to the difficulty in maintaining an intact seal on the test chamber and due to unloading the specimen prior to water testing. The test may not have accurately represented realistic field conditions because any potential cracks would have closed after the removal of the load [7].

One of the conclusions of this experimental investigation was that there was a significant difference between expected performance and observed performance. Although it appears to be a simple system, "Brick veneer attached to metal studs forms a complex load resisting mechanism. The complexity is caused by the difference in behavior of the two materials, which are constrained to act in parallel by ties that connect them" [9]. A review of the experimental testing performed in the early 1980s raised more questions than it answered and initiated additional analytical and experimental studies [4].

STUD HEIGHT

Stud height plays a direct role in the deflection of the wall system and the size of the cracks. The full-size testing performed at Clemson University and McMaster University [7,10] used wall heights of approximately 9 ft of brick veneer and 8 ft of cold-formed steel framing. The cold-formed steel studs in the Clemson tests were less than 8 ft in height because the specimens included a concrete spandrel element that served as the backing for a portion of the veneer. The full-size tests performed at the University of Alberta used wall heights of approximately 10 ft [11]. These heights are less than the typical floor-to-floor height for modern commercial and multifamily residential buildings (12.5 ft min.) and may not accurately represent the performance of a taller wall assembly.

TABLE 1 Summary of deflection recommendation history.

	Organization[a]					
	USG[b]	BIA[c]	ML/SFA (SSMA)[d]	NBC[e]	WSCPA[f]	AISI[g]
Period 1: Late 1960s to 1980	L/240 and L/360 studs only and composite wall; L/600 composite wall	L/600 to L/720 recommended; L/900 preferred; L/900 removed in 1980	L/360	L/360[h]		
Period 2: 1981 through 1985	Options based on BIA, Clemson, and USG testing	L/600 to L/720	L/360	L/360[h]		
Period 3: 1986 through 1989	Options based on BIA, Clemson, and USG testing	L/600	L/360[i]	L/360[h]		
Period 4: 1990 through 2002		L/600	L/240, L/360, and L/600 studs only[i]	L/600 including mech. play or L/720 with strict tie stiffness	L/360	
Period 5: 2003 to Current		L/600	L/240, L/360, and L/600 studs only	L/360	L/360	L/600

Note: [a]Shaded areas indicate that the organization did not have published guidelines during the period.
[b]United States Gypsum
[c]Brick Industry Association, previously Brick Institute of America
[d]Metal Lath/Steel Framing Association, now known as the Steel Stud Manufacturers Association
[e]National Building Code of Canada
[f]Western States Clay Products Association
[g]American Iron and Steel Institute
[h]References indicate that L/360 governed prior to this period, but first inclusion in code was not confirmed.
[i]Value is presumed.

Moisture

Over time, the concerns regarding moisture ingress have dissipated. Some of the performance failures in the early use of the BV/CFS system included widespread or persistent efflorescence and face spalling of brick [12]. Changes implemented early in the use of the system likely contributed to the reduction in reported efflorescence and face spalling over time, such as better specification of brick materials to ensure that only Grade SW (severe weathering) bricks were installed in BV/CFS systems [6]. Increased cavity widths and better workmanship to minimize mortar droppings in the air space permit increased airflow behind the brick to dry saturated areas and eliminate a source of soluble salts for efflorescence. As a result of these changes, efflorescence and face spalling are no more prevalent with this system than with other brick veneer wall systems.

The early versions of the BV/CFS system did not require a water-resistive barrier, leaving the sheathing, anchor/tie penetrations, and attachments to the studs exposed to moisture from bulk water or condensation [1]. Assumptions of rainscreen behavior were overly optimistic. General building science knowledge has increased over time, and relevant aspects of building science were included in the research conducted on this wall system. This increase in knowledge was applied to subsequent designs of BV/CFS wall assemblies. As a result, modern drainage cavities are better equipped to manage and expel any increased water from larger crack widths. Many of the materials currently used for air and moisture barriers (such as self-adhered rubberized asphalt membranes or liquid applied membranes) are robust enough to serve as a bulk water barrier for the gypsum sheathing, if detailed properly. Modern gypsum products also use fiberglass facings, eliminating the paper facings that served as a food source for biological growth in moist conditions. Some designers bypass the use of gypsum sheathing for cement board, which does not experience deterioration when exposed to moisture.

Cracking

Many of the arguments related to limiting the deflection of the BV/CFS system revolved around veneer cracking and limiting the size of those cracks. To date, no serviceability criteria or limits for crack sizes specific to brick masonry—much less brick veneer—have been established. As a result, many engineers have considered the crack limit of 0.015 in. established by the concrete industry for plain and reinforced concrete as an acceptable value for masonry cracks. This value is derived from the American Concrete Institute Committee Report, "ACI 224R, Table 4.1—Guide to Reasonable Crack Widths, Reinforced Concrete under Service Loads." The acceptable size of a crack in concrete is a function of cover and strain in the reinforcing steel and limiting the risk of corrosion. Interestingly, ACI 224R indicates that the tabulated crack widths are not considered an allowable upper bound of size and that doubling of crack width should be expected over several years [13].

This paper uses an existing simplified model to estimate cracking in brick veneer. It assumes a crack at mid-height, corresponding with the maximum wall deflection (Δv). Rigid body rotation of the veneer occurs around the crack like a hinge. Using similar triangles with the wall height, veneer thickness, and maximum wall deflection results in a maximum crack width (Δc). Refer to **Fig. 1**. Based on this model, to prevent cracking of the veneer, it is estimated to require a deflection in excess of L/2000 [14].

Given the typical dimensions of a single wythe brick veneer, restricting the crack size can result in excessively stringent deflection criteria. For instance, for a brick veneer with a specified width of 3 5/8 in., a deflection limit of L/3000 is needed in order to achieve a crack width of 0.005 in., which is considered barely visible (negligible). To achieve a crack width no wider than the referenced concrete crack limit of 0.015 in., a deflection limit of L/900 is needed. The deflection limit of L/600 to L/720 recommended by the BIA and confirmed by the McMaster University testing program was intended to limit crack widths to approximately 0.020 in. to 0.025 in., which is larger than the referenced concrete crack width recommendation but close enough in size to serve as a compromise between the crack width and deflection limit.

Other organizations recommend deflection limits of L/360 [15], which is consistent with structural roof members supporting plaster finishes and is more

FIG. 1 Schematic drawing for simplified calculation of crack width in brick veneer [11].

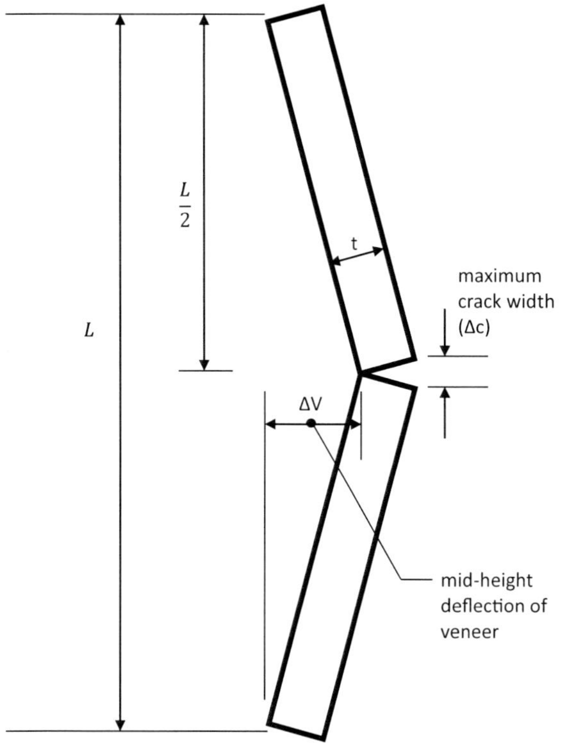

stringent than the typical deflection for walls with brittle finishes (L/240) per the building code. Using the same simplified model to estimate crack sizes, deflection of L/360 results in an estimated maximum crack width of 0.040 in. A crack width of this size is generally considered excessive because it is compared to the maximum concrete crack width of 0.015 in., particularly from the perspective of limiting moisture ingress; however, no empirical data exist to support or refute the potential consequence of this size crack. Cracks in the veneer caused by wind loads are not static. When a wind event ends, the cracks will close. Future wind events that are weaker than the design event will not reopen the cracks to the same degree. This dynamic nature of the crack width should be considered when trying to determine an acceptable upper bound value.

The Building Research Establishment in the UK surveyed more than 100 properties as part of a study on the effects of soil subsidence and established a system of damage categories associated with crack sizes called the Burland Scale. The six categories range from 0 to 5, with Category 0 as the smallest and Category 5 as the largest. Category 0 is for cracks up to 0.1 mm (0.004 in.), which are considered negligible. Category 1 is for cracks up to 1 mm (0.04 in.), which are considered fine and easily repaired. The Building Research Establishment considers cracks up to Category 2 (less than 5 mm or 0.20 in.) as aesthetic damage [16]. Although the Burland Scale was developed for permanent cracks, which differ from the type of dynamic cracks that develop within a BV/CFS system under wind loads, these categories can serve as an established benchmark to help determine an appropriate order of magnitude for acceptable crack widths, when they occur.

Wind Loads

Significant changes in the design wind loads have been incorporated over the years because they specified single value pressures based on building height. ASCE 7 acknowledged that the calculated design values, even if unfactored, were "excessively conservative" for the evaluation of serviceability and recommended reducing the calculated design wind load to determine the load for serviceability checks [17]. In early versions of ASCE 7, this was achieved by multiplying the basic wind speed design load by a reduction factor. Later, an alternate option developed. Currently, one option for reduced cladding loading is presented in the International Building Code (IBC) and the other is in the commentary of ASCE 7.

IBC METHOD

Footnote (f) in the deflection limit table of Section 1604 of the IBC allows for the reduction of the components and cladding wind load pressure for the purposes of calculating deflection. In IBC versions that reference ASCE 10 and later, the reduction factor in the footnote includes the multiplier to convert the strength design formula output to an unfactored load [18]. This method is similar to the original method proposed in ASCE 7.

ASCE 7 COMMENTARY METHOD

Prior to 2010, the ASCE 7 design loads were based on wind loads with a mean recurrence interval (MRI) of 50 years. Even though the current ASCE 7 design loads are based on wind events with MRI 700 years for typical Risk II structures, the resulting unfactored loads approximate those in ASCE 7-2005, as shown in Table 2. The ASCE 7 commentary recommends the use of alternate wind speed maps for events with MRI 10 years to 100 years to determine loads for serviceability in lieu of a multiplier. It leaves the decision of which MRI wind event to use in serviceability design up to the engineer's judgment and consultation with the building owner to determine the desired serviceability performance [19].

Many buildings have an intended service life of 30 to 60 years; therefore, a wind event with a low MRI (such as 10 years) [20] typically is appropriate for deflection calculations given that the event could be expected several times over the service life of the building. In all cases, anchors/ties would be adequately sized and spaced in accordance with full design loads from a strength perspective.

Anchors/Ties

MECHANICAL PLAY AND DEFORMATION

The anchors/ties connecting brick veneer to CFS studs were initially assumed to be sufficiently rigid to transfer load directly to the studs; however, testing demonstrated that their behavior was more complex [7]. Prescriptive requirements for veneer anchors are limited, which permits a wide variety of anchor/tie types and configurations. "TMS 402—Building Code Requirements for Masonry Structures" [5] includes only minimum wire sizes and plate thicknesses. Adjustable anchors/ties consist of two parts that are able to translate in relation to each other to accommodate movement in the brick veneer. By design, the pieces do not fit together rigidly to permit this movement, and there is mechanical play within the anchor/tie. Although of adequate stiffness to transfer load among the wythes, many anchor

TABLE 2 Design wind load summary.

Code	UBC	UBC	ANSI	ASCE	ASCE	ASCE	ASCE	ASCE	ASCE	ASCE	ASCE
Year	1967	1982	1982	1988	1993	1995	1998	2002	2005	2010	2016
Max load direct from formulas (psf)	30	34	31	31	31	42	34	35	34	58	55
ASD load combination coefficient	-	-	-	1.0	1.0	1.0	1.0	1.0	1.0	0.6	0.6
Unfactored design load (psf)	30	34	31	31	31	42	34	35	34	35	33

Note: All loads assume Zone 5 (negative) at the top of the building (110 ft), with an assumed area of 30 ft^2 (tributary stud area—up to 24-in. spacing and 15 ft tall). Components and cladding loads unless no distinction provided. UBC = Uniform Building Code; ANSI = American National Standards Institute.

and tie types consist of relatively thin plates and wires, which can be subject to deformation under axial load. Early models idealized the anchor/tie as a link element and did not account for movement or deformation in the manufactured product. Recommendations limiting the amount of mechanical play and deformation under a given load were developed by the BIA [21], with the mechanical play provision incorporated into TMS 402 in 1995 when veneer requirements were added.

When these values of mechanical play and deformation are incorporated into the calculation of the deflection and the maximum crack width for the system, the deflection of the veneer exceeds that of the cold-formed steel backing. The magnitude of the mechanical play (only half is considered in calculations) and the anchor/tie deformation is a significant component. As shown via a simplified analysis in **Table 3**, stud selection based on an L/600 deflection correlates with an estimated veneer deflection between L/360 and L/480. It appears the more stringent deflection requirement for the cold-formed steel backing accommodated the additional flexibility of the anchors/ties within the assembly, which was not explicitly understood at the time.

TABLE 3 Veneer deflection and estimated maximum crack widths.

L (ft)	Stud Criteria (L/x)	Stud Defl. (in.)	Mechanical Play (in.)	Deform. at 100 lb (in.)	Total Midpoint Deflection (in.)	Est. Max. Crack Width (in.)	Veneer Defl. (L/x)
8.5	360	0.283	0.0625	0.05	0.365	0.052	280
9	360	0.300	0.0625	0.05	0.381	0.051	283
9.5	360	0.317	0.0625	0.05	0.398	0.051	286
10	360	0.333	0.0625	0.05	0.415	0.050	289
14	360	0.467	0.0625	0.05	0.548	0.047	307
14.5	360	0.483	0.0625	0.05	0.565	0.047	308
15	360	0.500	0.0625	0.05	0.581	0.047	310
8.5	480	0.213	0.0625	0.05	0.294	0.042	347
9	480	0.225	0.0625	0.05	0.306	0.041	353
9.5	480	0.238	0.0625	0.05	0.319	0.041	358
10	480	0.250	0.0625	0.05	0.331	0.040	362
14	480	0.350	0.0625	0.05	0.431	0.037	390
14.5	480	0.363	0.0625	0.05	0.444	0.037	392
15	480	0.375	0.0625	0.05	0.456	0.037	395
8.5	600	0.170	0.0625	0.05	0.251	0.036	406
9	600	0.180	0.0625	0.05	0.261	0.035	413
9.5	600	0.190	0.0625	0.05	0.271	0.035	420
10	600	0.200	0.0625	0.05	0.281	0.034	427
14	600	0.280	0.0625	0.05	0.361	0.031	465
14.5	600	0.290	0.0625	0.05	0.371	0.031	469
15	600	0.300	0.0625	0.05	0.381	0.031	472

Because there is a wide variety of combinations for mechanical play and deformation, the maximum values for these characteristics were used in the table. The influence of the anchor/tie deformation and mechanical play is less pronounced for taller walls because the maximum mechanical play and deformation is constant as the length increases.

Both veneer deflection and maximum estimated crack width can be reduced if either the anchor mechanical play or the anchor deformation (or both) are less than the maximum allowable values. Product literature for veneer anchors in the United States generally does not include any information about the mechanical play or deformation of that anchor, even though these parameters need to be considered by the designer to meet code. From 1995 through 2005, Canadian code requirements for anchor mechanical play and deformation were more stringent than those in the United States [22] and, consequently, Canadian product literature includes these values to permit designers to compare values during the product selection phase of the design process. Standardized test procedures developed to evaluate these limits for anchors/ties are not currently established in the United States but can be developed based on the tests used in Canada and on similar experimental investigations conducted in the United States. Part 4 of the McMaster University testing program [23] recommended the development of performance indexes that could help designers sort through the variations in mechanical play and deformation, especially when anchors are at the limits of their adjustability.

Allowable Crack Width

As presented in **Table 3**, for stud heights of many common wall configurations (8.5 ft to 15 ft), the estimated maximum crack width ranges from approximately 0.03 in. to 0.05 in. based on stud deflections of L/600 to L/360. If one applies the Burland Scale as discussed previously, these crack widths would be classified as falling within the upper range (seventy-fifth percentile or higher) of Category 1 or a much lower range (tenth percentile or lower) of Category 2. Cracks in Categories 1 and 2 are considered aesthetic, indicating that they should be acceptable from an aesthetic perspective.

Veneer cracks only reach the calculated maximum width during a load event in which the deflection of the stud reaches the serviceability limit (stud criteria per **Table 3**). As discussed previously, the frequency and intensity of these events vary such that, during typical wind events, the crack width will be smaller than the calculated maximum. On a daily basis, when winds are negligible, the crack may be virtually imperceptible.

Minimizing veneer cracks to limit the amount of water penetrating the backing is no longer one of the primary means of controlling water entry to the interior of the building. Use of contemporary methods and improved materials in the drainage cavity has resulted in less reliance on veneer crack control.

All things considered, a maximum veneer crack width of approximately 0.05 in. is acceptable from both an aesthetic and a moisture perspective.

Method 1: Maximum Crack Size

If we assume that a nominal width of 0.05 in. is an acceptable maximum veneer crack limit, many of the disagreements regarding deflection become moot. Setting a crack size as the design criteria permits the use of multiple combinations of stud deflection and anchor play/deformation to achieve a given result. This method permits the designer to meet the desired performance criteria (crack size) without the restriction of a particular deflection limit that may not be consistent with the desired performance criteria.

The width of the veneer also contributes to the maximum crack width, so variation here can also be beneficial. Brick with specified widths of 3.5 in. and 3 5/8 in. are both considered 4-in. nominal units, but the 3.5-in. unit results in a narrower estimated crack width.

Considering the veneer crack width as a serviceability limit will add design steps and decisions to the cold-formed steel stud selection process, but the overall approach is similar to current BV/CFS design practice. The following procedure is idealized and assumes equal deformation of the veneer sections and limited movement at the top and bottom of the veneer. However, note that experimental testing has shown that significant movement can occur at these locations, which results in smaller veneer cracks than the projected crack width.

1. Select the maximum allowable crack width for the application (Δc).
2. Obtain or select the specified brick dimension (w).
3. Use Eq 1 to determine the maximum midpoint deflection of the system (Δv), where L is the height of the wall.

$$\Delta c = \frac{4 \cdot \Delta v \cdot w}{L} \quad (1)$$

4. Subtract the anchor/tie flexibility component from Δv: one-half the mechanical play plus the deformation of the anchor under a 100 lb load.
5. Use the remaining deflection value to determine the minimum moment of inertia needed for the cold-formed steel stud.
6. Select cold-formed steel stud from manufacturer tables based on effective inertia.

Method 2: Maximum Stud Deflection

This method uses the typical existing procedure for stud selection familiar to most designers. Once the minimum moment of inertia and stud size are determined, proceed as follows:

1. Determine actual stud deflection for given load case.
2. Obtain mechanical play and deformation information for the anchors/ties under consideration.
3. Calculate estimated maximum crack width using simplified analysis and compare to maximum allowable crack width of 0.05 in.

By using the crack width as the serviceability parameter, designers now have the ability to select a stud deflection tailored to the specific project requirements anywhere within the range of L/600 to L/360. This permits accommodation of variable deflection limits depending on the desired level of performance.

Theory versus Practice

Because BV/CFS wall systems have been designed using multiple methods, it would be beneficial to better understand how the different recommendations have been implemented over the years. Arumala, in 2007 [24], speaks to an informal understanding that low-rise buildings typically are designed to L/360 deflection, and taller buildings are designed to L/600 deflection, which mirrors the recommendations of the Western States Clay Products Association for two tiers or grades of construction. There have been approximately ten years of construction in Canada since the deflection requirements for brick veneer on CFS were reduced, and it would be instructive to understand the results of such a change. To our knowledge, there have not been high profile failures of systems designed using the reduced deflection requirements. But, like the Canada Mortgage and Housing Corporation study in 1986 [25] was successful in identifying the need for more knowledge of the system, a newer study would likely be able to ascertain whether deflection criteria correlate with performance in the field. Anecdotally, overall performance of the BV/CFS system has improved significantly through the years, but specific objective data on wall deflections and cracking are lacking.

Outstanding Issues

To date, much of the concern about cracking has been related to the potential for such cracks to permit moisture to enter the wall system, even though the system is a drained assembly. Deterioration over time to both cracked mortar joints and anchor engagement was once raised as a potential issue [4], but this issue was never pursued directly in an experimental study. Testing of full-size wall specimens demonstrated that failure of adjustable anchor components or local buckling of the CFS studs occurred prior to disengagement of the anchor from the veneer bed joint, suggesting that veneer cracking may not significantly reduce the anchor/veneer strength [10,11].

However, the testing performed to date has not replicated the quantity of cyclic loads and bed joint cracking that could occur under repeated events over the service life of a veneer on a typical building. Where both anchor/tie locations and crack locations were documented, the cracks and anchors/ties only coincided in one case [11], but it is not clear if that result is typical. There were no reports of anchor/tie disengagement from the veneer during the testing. Experimental studies to investigate the maximum veneer crack width limits are recommended to definitively establish cracking limits.

Recommendations

In summary, the following recommendations are proposed:

1. Consider using an allowable maximum crack size of 0.05 in. in the veneer as an alternate criterion to a deflection of L/600 applied to the cold-formed steel.

2. Encourage anchor/tie manufacturers to provide information on the mechanical play and deformation in their published literature. Standardization of testing may be required.
3. Conduct a survey, similar to that performed by the Canada Mortgage and Housing Corporation in 1986, to identify design practices and potential performance differences in the United States and Canada.
4. Develop a future testing program to study the behavior of the brick veneer itself for use in refining the serviceability recommendations.

References

[1] Ramirez, I., "Structural Performance of Brick Veneer/Steel Stud Wall Systems as Designed in Current Practice," M.S. thesis, University of Texas at Austin, Austin, TX, 1988.

[2] Brick Institute of America, "Brick Veneer Panel and Curtain Walls—28B," *Technical Notes on Brick Construction*, Brick Institute of America, Reston, VA, 1979.

[3] Clark, C. B., Jr., Fried, C. L., and Bryja, J., "History of Cold-Formed Steel Framing Used as a Backing for Brick Veneer," *Masonry 2018, ASTM STP1612*, N. V. Krogstad and W. M. McGinley, Eds., ASTM International, West Conshokocken, PA, 2018, pp. XX–XX, http://dx.doi.org/10.1520/STP1612/20170012

[4] Bell, G. R. and Gumpertz, W. H., "Engineering Evaluation of Brick Veneer/Steel Stud Walls, Part II—Structural Design, Structural Behavior, and Durability," *Proceedings of the Third North American Masonry Conference*, Arlington, TX, June 3–5, 1985, J. H. Matthys and J. G. Borchelt, Eds., University of Texas at Arlington, Arlington, TX, 1985, pp. 5-1–5-20.

[5] The Masonry Society, "TMS 402—Building Code Requirements for Masonry Structures," The Masonry Society, Longmont, CO, 2016.

[6] Brick Institute of America, "Brick Veneer Panel and Curtain Walls—28B," *Technical Notes on Brick Construction*, Brick Institute of America, McLean, VA, 1980.

[7] Arumala, J. O. and Brown, R. H., "Final Report—Performance Evaluation of Brick Veneer with Steel Stud Backup," Department of Civil Engineering, Clemson University, Clemson, SC, 1982.

[8] Metal Lath/Steel Framing Association; Brick Institute of America, "Joint Statement on the Clemson University Brick Veneer-Steel Stud Testing Program," Brick Institute of America, McLean, VA, 1982.

[9] Kelly, T., Goodson, M., Ronald, M. and Asher, J., "Analysis of the Behavior of Anchored Brick Veneer on Metal Stud Systems Subjected to Wind and Earthquake Forces," *Proceedings of the Fifth North American Masonry Conference*, University of Illinois at Urbana-Champaign, June 3–6, 1990, D. P. Abrams, Ed., Masonry Society, Boulder, CO, 1990, pp. 1359–1370.

[10] Drysdale, R. and Wilson, M., "Part 5—Tests of Full Scale Brick Veneer Steel Stud Walls to Determine Strength and Rain Penetration Characteristics," Canada Mortgage and Housing Corporation, Ottawa, ON, 1990.

[11] McGinley, W., Warwaruk, W., Longworth, J. and Hatzinikolas, M., "Stuctural Engineering Report 156—Masonry Wall Veneer Systems," Department of Civil Engineering, University of Alberta, Edmonton, AB, 1988.

[12] Drysdale, R. G. and Suter, G. T., "Exterior Wall Construction in High-Rise Buildings—Brick Veneer on Concrete Masonry or Steel Stud Wall Systems," Canada Mortgage and Housing Corporation, Ottawa, ON, 1991.

[13] American Concrete Institute, "ACI 224R—Control of Cracking in Concrete Structures," American Concrete Institute, Farmington Hills, MI, 2001.

[14] Structural Clay Products Research Foundation, "Research Report Number 9—Compressive, Transverse, and Raking Strength Tests of Four-Inch Brick Walls," Geneva, IL, 1965.

[15] KPFF Consulting Engineers, "Design Guide for Anchored Brick Veneer Over Steel Studs," Western States Clay Products Association, Orinda, CA, 1995.

[16] Borough of Camden, "Camden Planning Guide—Basements and Lightwells CPG4," Borough of Camden, London, July 2015.

[17] American Society of Civil Engineers, "ASCE 7—Minimum Design Loads for Buildings and Other Structures," American Society of Civil Engineers, New York, 1995.

[18] International Code Council, "International Building Code," International Code Council, Country Club Hills, IL, 2015.

[19] American Society of Civil Engineers, "ASCE 7—Minimum Design Loads for Buildings and Other Structures," American Society of Civil Engineers, Reston, VA, 2010.

[20] Griffis, L. G., "Serviceability Limit States Under Wind Load," *Engineering Journal*, American Institute of Steel Construction, Vol. 30, 1993, pp. 1–16.

[21] Brick Institute of America, "Wall Ties for Brick Masonry—44B," *Technical Notes on Brick Construction*, Brick Institute of America, McLean, VA, 1987.

[22] Canadian Sheet Steel Building Institute, "Lightweight Steel Framing Architectural Design Guide," Canadian Sheet Steel Building Institute, Cambridge, ON, 2002.

[23] Drysdale, R. and Wilson, M., "Part 4—Behaviour of Brick Veneer/Steel Stud Tie Systems," Canada Mortgage and Housing Corporation, Ottawa, ON, 1989.

[24] Arumala, J. O., "Brick Veneer Steel Stud Wall Systems: State-of-the-Art," *TMS Journal*, Vol. 25, No. 1, 2007, pp. 9–19.

[25] Suter Keller, Inc., "Brick Veneer/Steel Stud Wall Design and Construction Practice in Canada—Results of a 1986 Survey," Canada Mortgage and Housing Corporation, Ottawa, ON, 1986.

Works Consulted

American Iron and Steel Institute, "Brick Veneer Cold-Formed Framing Design Guide—D112," American Iron and Steel Institute, Washington, DC, 2013.

American Iron and Steel Institute, "Steel Stud Brick Veneer Design Guide—CF03-1," American Iron and Steel Institute, Washington, DC, 2003.

BRE Group, "Assessing Cracks in Houses," http://web.archive.org/web/20180208211908/https://www.bre.co.uk/page.jsp?id=3356 (accessed August 28, 2017).

Brick Industry Association, "Brick Veneer: Brick Veneer/Steel Stud Walls—28B," *Technical Notes on Brick Construction*, Brick Industry Association, Reston, VA, 1999.

Brick Industry Association, "Brick Veneer/Steel Stud Walls—28B," *Technical Notes on Brick Construction*, Brick Industry Association, Reston, VA, 2005.

Brick Institute of America, "Brick Veneer Steel Stud Panel Walls—28B," *Technical Notes on Brick Construction*, Brick Institute of America, McLean, VA, 1987.

Brown, R. H. and Arumala, J. O., "An Experimental Study of Brick Veneer with Metal Stud Backup," *Proceedings of the Sixth International Brick and Block Masonry Conference*, Rome, Italy, May 16–19, 1982, ANDIL, Rome, pp. 1411–1427.

Building Research Establishment, "BRE Digest 251—Assessment of Damage in Low-Rise Buildings," Building Research Establishment, Garston, Watford, UK, 1995.

Canada Mortgage and Housing Corporation, "Best Practice Guide—Building Technology—Brick Veneer Steel Stud," Ottawa, ON, 1996.

Canadian Sheet Steel Building Institute, "Steel Studs as Back-up for Brick Veneer Changes in NBC 2005," *Lightweight Steel Framing Technical Bulletin*, Canadian Sheet Steel Building Institute, Cambridge, ON, 2006.

Chin, I. R., Krogstad, N. V., and Monk, C. B., "Influence of Tie Flexibility, Relative Length, and End-Boundary Condition on Brick Veneer-Metal Stud Flexural Bond Stress," *Masonry: Materials, Design, Construction, and Maintenance ASTM STP992*, H. A. Harris, Ed., ASTM International, West Conshohocken, PA, 1988, pp. 96–117, http://dx.doi.org/10.1520/STP992-EB

Drysdale, R. G. and Breton, N., "Part 1—Strength and Stiffness Characteristics of Steel Stud Backup Walls Designed to Support Brick Veneer," Department of Civil Engineering and Engineering Mechanics, McMaster University, Hamilton, ON, 1991.

Drysdale, R. G. and Kluge, A., "Part 3—Performance of Brick Veneer/Steel Stud Wall Systems Subject to Temperature, Air Pressure, and Vapour Pressure Differentials," Canada Mortgage and Housing Corporation, Ottawa, ON, 1990.

Ghosh, S. K., "Searching for Simplicity: The Evolution of Wind Provisions in Standards and Codes in the United States," *RCI Interface*, Vol. 26, 2008, pp. 13–21.

Gumpertz, W. H. and Bell, G. R., "Engineering Evaluation of Brick Veneer/Steel Stud Walls, Part I—Flashing and Waterproofing," *Proceedings of the Third North American Masonry Conference*, Arlington, TX, June 3–5, 1985, J. H. Matthys and J. G. Borchelt, Eds., The Masonry Society, Longmont, CO, 1985, pp. 4-1–4-16.

Laird, D. A., Drysdale, R. G., Stubbs, D. W., and Sturgeon, G. R., "The New CSA S304.1-04—Design of Masonry Structures," *Proceedings of the Tenth Canadian Masonry Symposium*, Banff, Alberta, June 8–12, 2005, S. Lissel, Ed., University of Calgary, Department of Civil Engineering, Calgary, AB, 2005, pp. 312–321.

Marchello, M. J., "Steel Stud Curtain Wall Design Stands Up to Rigorous Tests," *Form & Function*, USG Corporation, Chicago, IL, 1984.

Marchello, M. J., "U.S.G. Tests Provide New Data for Determining Limiting Heights of Steel-Stud/Brick-Veneer Curtain Wall Assemblies," *Form & Function*, Issue 3, USG Corporation, Chicago, IL, 1981, pp. 14–15.

McGinley, W. M., Longworth, J., Warwaruk, J., and Hatzinikolas, M., "Masonry Veneer and Steel Stud Curtain Walls," *Proceedings of the 4th Canadian Masonry Symposium*, Fredericton, New Brunswick, June 2–4, 1986, J. L. Dawe, R. M. Francis, and A. J. Valsangkar, Eds., University of New Brunswick, Department of Civil Engineering, Fredericton, New Brunswick, 1986, pp. 730–743.

Mehta, K. C., "Wind Load History: ANSI A58.1-1972 to ASCE 7-05," *Proceedings of the 2010 Structures Congress*, Orlando, FL, May 12–15, 2010, S. Senapathi, K. Casey, and M. Holt, Eds., ASCE, Reston, VA, 2010, pp. 2134–2140.

National Concrete Masonry Association, "Laboratory Investigation of Metal Stud Back Up Systems for Masonry Veneers," National Concrete Masonry Association, Herndon, VA, 1981.

Postma, M. A. and Burnett, E. F. P., "Renovation Strategies for Brick Veneer Steel Stud Wall Construction—Task 2," University of Waterloo for Canada Mortgage and Housing Corporation, Waterloo, ON, 1993.

Swensson, K. D., "Guide to the Design of Building Systems for Serviceability in Accordance with the 2012 IBC and ASCE/SEI 7-10," International Code Council, Country Club Hills, IL, 2013.

West, M., Fisher, J., and Griffis, L., "AISC Steel Design Guide 3—Serviceability Design Considerations for Steel Buildings," American Institute of Steel Construction, Washington, DC, 2003.